S&AC 230501

Entropy and Air Conditioning
An Historical Introduction

By Robert Turtle

Copyright © 2023
Robert R. Turtle

This ebook may not be sold or given away to other people. If you would like to share this book with another person, please purchase an additional copy for each recipient. If you're reading this book and you did not purchase it, or it was not purchased for your use only, then please return to the ebook retailer and purchase your own copy. Thank you for respecting the hard work of this author.

Table of Contents

About the Cover
Chapter 1--Keeping Cool with MKS Units
Chapter 2--Temperature
Chapter 3--Heat and How It Gets Around
Chapter 4--The Mechanical Equivalent of Heat
Chapter 5--The First Law of Thermodynamics
Chapter 6--Heat Transfer
Chapter 7--Calculating with Ideal Gases
Chapter 8--Air Cycle Air Conditioner
Chapter 9--Children of Prometheus
Chapter 10--Carnot's Engine and the Second Law of Thermodynamics
Chapter 11--Clausius' Entropy
Chapter 12--Enthalpy Change in Refrigeration
Chapter 13--Phase Changes
Chapter 14--What is Entropy?
Chapter 15--Cooling with Nozzles and Throttles

Chapter 16--Two Phase Refrigeration Cycle
Works Cited

About the Cover

Why would anyone write a book with a title and cover such as you see here? It is impossible to contemplate the cost of replacing an aging residential air conditioning system without sensations of anxiety. Promotional literature is filled with a strange blend of units and acronyms and may do little to instill confidence. This book will show that the most straightforward, low-tech approach to home air conditioning would lead to a system that is impractically bulky. To describe and explain the systems currently in use, it is necessary to introduce concepts from advanced college courses in thermodynamics and chemistry. Among these, entropy seems both essential and deeply mysterious.

Like pressure, volume, and temperature, entropy S is a property of matter. Unfortunately, there appears to be nothing like a thermometer to measure it directly--the entropy change of a sample is found as the amount of heat it absorbs divided by the absolute temperature provided the temperature change is small.

The cover image shows how to calculate the entropy change of a sample of ideal gas due to small relative changes in the absolute temperature T and volume V. The method is introduced in Chapter 7 and applied later in Chapter 14 when Boltzmann's theoretical model for entropy is discussed. It is motivated by the recognition that entropy is a state variable, so its change can be calculated along the most convenient path between two points in thermodynamic variable space.

Chapter 1
Keeping Cool with MKS Units

A gentle breeze of cool, dry air on even the hottest summer day is what we rely on air conditioning to supply. A fan vigorously stirring warm, muggy air is better than nothing, but a strong breeze moves papers and can make cooking on a gas stove difficult. Thoughts turn to ice, which might do the job, for melting ice cools our drinks and is useful for preserving food on car trips.

A fan blowing air past chunks of ice might cool a room without increasing humidity the way an evaporative swamp cooler would, especially if the ice and melt water were kept in plastic bags or in some way that prevented water from evaporating. There is a performance margin, since ice is able to cool to near the freezing point of water, 32 F or 0 C, while the goal for room temperature is higher, perhaps as high as 77 F or 25 C. This apparent temperature mismatch can help drive thermal transport, so water ice is potentially useful for air conditioning even though it seems better suited for food storage. Air conditioner performance is often specified by the number of tons of water a unit could freeze to ice in twenty-four hours. (Ananthanarayanan, p. 7) This traditional rating scheme does not encourage bringing ice home from the grocery store and using it to cool a home, since a nominal capacity of three to four tons may be provided for cooling a two-bedroom house in the American Southwest. However, it has been suggested that ice making could be a means for load leveling when rooftop solar power is used to power air conditioners. This approach would bypass reliance on electric storage batteries.

The name for a unit of weight or mass seems misapplied to naming a cooling rate. More significantly, about 3.52 kilowatts (kW) of electric power would be needed to drive the current in a heating coil that could melt a ton of ice in a day, yet a "one ton" air conditioner may draw only a little over a kilowatt of power when operating. What has happened to the conservation of energy? What seems too good to be true turns out to be a beneficial consequence of the second law of thermodynamics, which will be taken up later. It is not part of a cheating scheme. The electrical power drawn for air conditioning can be significantly less than what an estimate based directly on the heat of fusion for ice suggests.

Less than a century ago, the amount of ice a machine could produce in a day was a practical quantity, because city dwellers could order blocks of ice to be delivered regularly to their kitchen for use in "iceboxes." At the time, the name was literally valid, and the gentlemen who delivered had special vests to protect their shoulders and scary-looking tongs for grabbing glistening, slippery blocks that seemed immense to a small child witnessing the process.

Artificial refrigeration has been in widespread general use for little more than a century. This is partly because ice can be collected in winter and saved through summer in cellars. It was sometimes transported considerable distances in ships. (Williams, 2012) However, artificial refrigerants that will cool to freezing temperature when fanned with air have been known at least since the synthesis of ether from alcohol and sulfuric acid was discovered. (Youmans, p. 305) Various means were found to produce artificial cooling. (Worthing, p. 364) By 1834 Jacob Perkins, a United States inventor, had developed a successful machine based on evaporating ether, which was condensed and reused as part of the process. (Sandfort, p. 164) Ether is a hazardous substance, and we seldom encounter it in our lives today. However, it is relatively easy to synthesize, and it was once widely used for important purposes before better and safer substitutes were found.

Compressing air, cooling it back toward ambient temperature, and then allowing it to expand against a piston in a cylinder is an alternate approach for producing low temperatures. It was developed in the 1840s by another United States inventor, Dr. John Gorrie. He worked in Florida, where ice was sorely needed in summer (Sandfort, p. 166).

Drawings of Perkins' and Gorrie's machines suggest they were able to take advantage of steam engine design practice, which was advancing rapidly in the dawning age of railroads. Gorrie's air cycle approach is seldom used for refrigeration today, but a similar thermodynamic cycle, the Brayton cycle, is alive and well in turbojet engines that power aircraft.

Perkins' refrigerator, the ancestor of present-day air conditioners and refrigerators, used a thermodynamic cycle similar to that of today's equipment. Inflammable, toxic ether has been replaced, but the ideal choice of refrigerant remains unclear partly because of the need to protect the earth's ozone layer. (Masterton, p. 666) Later we will work examples using data for ammonia, a refrigerant that is still widely used despite its dangerous properties.

Like the horsepower, the ton unit has both a literal and a historical significance that help keep its memory alive. But here we will stick almost entirely to what are informally called mks units because they are based on the meter length, the kilogram mass, and the second of time. This helps to stay in touch with the parallel insights that engineering, chemistry, and other branches of science can provide.

We will need to pick a temperature unit to use consistently, too. In the Southwestern United States, people start from seeing and hearing temperatures in degrees Fahrenheit on the news and must move on in two steps to think about temperatures on the Kelvin scale, which is the best single choice here. Kelvin temperatures are always greater than zero, and they are indicated by the symbol K. Here are some significant temperatures for air conditioning:

Liquid ammonia (a practical refrigerant) boils in an open container at about 240 K.

Water freezes at 273 K.

Unwanted ice formation can be a problem since 273 K is only a little colder than the temperatures at which refrigerants are expected to evaporate in air conditioner heat exchangers. This is most likely to be a problem where humidity is high.

The temperature of conditioned air emerging from a cooling duct could be 286 K.

The corresponding indoor temperature set on a thermostat could be 298 K.

Normal body temperature is about 310 K.

The outdoor temperature on a summer day in southern Nevada can be 314 K or higher.

Be it ever so hot outside, the refrigerant vapor must be compressed adiabatically until it can condense at a temperature that is hotter still.

Water boils at 373 K at sea level.

Other sources may use different values for what we are calling reference temperatures. One could argue that the goal room ambient temperature suggested here, 298 K, is on the warm side.

We're talking about temperature when we say something is hot or cold. Heat is something else. It spreads naturally from hotter objects to colder ones, and the hotter objects grow colder and the colder objects grow hotter.

Temperature differences tend to die out, and the resulting intermediate temperature can persist until a new source or sink for heat is unleashed. The process often involves direct contact between substances, but heat can be transferred by electromagnetic waves even through vacuum. This is how we receive energy from the sun, and the earth remains at a steady average temperature because it emits energy into space in the form of infrared radiation. Many artificial satellites and spacecraft rely on radiant energy transfer for temperature control, and keeping temperatures within specified limits can be an engineering challenge.

From the point of view here, it seems generally correct to say that heat is energy associated with random molecular movements even though it may be spread by electromagnetic radiation. (Youmans, p. 61)

Temperature and heat are partly under our control. We light fires to obtain a source of heat, and we use blankets to prevent heat from dissipating.

Heat eventually penetrates insulation where there is a temperature gradient, and temperature differences die out when sources of heat are eliminated and things are left to themselves. There is a large range of thermal phenomena to deal with. One that is especially important here is the way temperature is stabilized by a phase change even while heat conduction is taking place. Ice melting to water at about 273 K is perhaps the most familiar example and has already been mentioned. Liquid refrigerant evaporating to gas in a heat exchanger coil is another.

After the distinction between temperature and heat was grasped in the eighteenth century, the next big advance in understanding was to recognize that the flow of heat amounts to a transfer of energy in a form comparable to the work one does walking up a hill or raising a load with a block and tackle. A milestone was passed when American expatriate Benjamin Thompson realized that mechanical friction could produce unlimited quantities of heat for as long as mechanical energy was provided to keep the process going. (Sears, p. 270)

Chapter 2
Temperature

Our sense of touch tells us what is hot and what is cold in a way adapted to everyday needs. This sense can be quite subjective. Once while running water into an unusual, high-end bathtub equipped with a settable tap temperature, the writer was surprised by a relatively abrupt transition from what felt cool to what felt hot when the dial setting was moved a short way past body temperature at 37 C. This is a familiar type of experience. Swimmers often find water cold at first; however, they grow accustomed to it, and what was unpleasantly chilly soon becomes more comfortable. (Youmans, p. 61) Thermometers take this sort of subjectivity out of temperature descriptions.

Cooking benefits from using a thermometer. One can tell whether a ready-made quiche from the grocery store has been heated hot enough to kill bacteria without waiting for it to bubble in the microwave. No clear notion of exactly what the thermometer measures is needed in such circumstances. However, so many phenomena are influenced by temperature that it is natural to ask what exactly a thermometer is responding to. Understanding of the relation between temperature and the flow of heat evolved gradually.

In the seventeenth century, the properties of air were being investigated with the help of transparent glass tubes partly filled with liquids. Robert Boyle discovered the inverse relationship between pressure and volume for air at constant temperature, and the variation of atmospheric pressure with altitude was measured with the mercury barometer. During the eighteenth century, information was accumulating to suggest that gas pressure might be due to molecular motion. The French scientist Amontons invented the concept of absolute zero when he extrapolated temperature measurements made with an air thermometer to zero pressure. (Garriga y Buach, p. 167) This observation led to the idea that the absolute temperature (roughly speaking the Kelvin temperature) might be proportional to the average kinetic energy of molecules in a gas sample.

Later in the eighteenth century, Jacques Charles and Joseph Gay-Lussac built on the work of Amontons. Gay-Lussac found air increases in volume by a factor of about 1.37 between the freezing and boiling points of water independent of the pressure as long as it remained the same throughout the expansion. (Garriga y Buach, p. 168) Their work led to the familiar gas law

$$PV = RT$$

that relates pressure and volume to temperature measured relative to Amontons' zero point. This relationship suggests that if one knows the value of any two of the three variables, then one can calculate the value of the third. On today's familiar Kelvin scale of absolute temperature, the freezing point of water is 273 K, the boiling point at sea level is 373 K, and 373/273 = 1.366.

At the start of the nineteenth century, the flow of heat from hot to cold was explained by the idea that heat might be an invisible fluid that flowed naturally from high temperature to low more or less the way water wets a paper napkin. Like the water in the napkin, the amount of heat remains constant as the temperature distribution changes with time. This is known as the caloric hypothesis, and it once gained wide acceptance. (Garriga y Buach, Cap. II)

Even though the concept of kinetic energy was not yet widely applied in physics, Benjamin Thompson suggested that increasing temperature indicates increasing agitation at the molecular level. People were not prepared to work with this idea at first. Thompson was onto something, however, when he noted how friction seemed capable of producing any amount of heat desired.

At the start of the nineteenth century, work calculated as force times distance had come into its own as both a measurable and an economic quantity, and carefully measuring the temperature change in a sample of material that was produced by a given amount of work expended as friction eventually showed the way out of the dilemma introduced by Benjamin Thompson's observations. By 1875, textbooks were treating heat as molecular agitation in terms that are still familiar today. (Youmans, p. 63)

A plastic pill bottle and a length of small diameter plastic tubing can be used to demonstrate the thermal properties of air. A snug hole for the tube is made in the bottle cap, and the juncture between the hole in the cap and the tube wall is sealed with petrolatum, which is also known as petroleum jelly and distributed under a familiar trade name. The seating area of the cap also needs a coating. When the tube has been partly filled with water, the cap can be screwed onto the bottle. Then the water will move back and forth in the tube in response to warming the bottle by hand and then allowing it to cool. A colored bottle will quickly warm up in sunlight.

(It is worth noting that this is approximately a constant volume gas thermometer where the liquid levels change in response to pressure changes. If the liquid reaches the open end of the tube, it will all quickly run out onto the floor.)

An air thermometer is sensitive, but it is difficult to calibrate and not very portable. Bimetallic strips that flex as the temperature changes are useful, but the most recognizable form of thermometer is the glass bulb thermometer.

Glass bulb thermometers had been in use for decades before Charles and Gay-Lussac quantified the thermal properties of air by introducing the law bearing their name. The familiar liquid-in-glass thermometer is made entirely of glass and has a capillary extension tube so the larger expansion and contraction of a liquid relative to the solid glass enclosure can be observed. In a noteworthy technical tour-de-force, the liquid is sealed in the tube and completely surrounded by glass. This form of thermometer made reproducible measurements easy to carry out. Reference temperatures (especially the boiling and and freezing points of water) were identified so measurements by different observers with different thermometers could be compared. The dependence of the boiling point on altitude was recognized, and calibration methods took this into account.

A glass thermometer with a bulb and slender tube that is left open at the top can be filled with liquid and used to sense temperature change. This is a practical configuration, but not as stable as the one with the glass tube sealed off. Apparently Daniel Fahrenheit improved on earlier work and then went on to manufacture sealed liquid-in-glass thermometers on a commercial scale.

A notable feature of the liquid in glass thermometer is that the sensing volume of alcohol or mercury can be less than a cubic centimeter. Temperatures can be read in many situations, since the thermometer does not need to be fully immersed, and it can be used tipped at an angle.

Even though his temperature scale does not enjoy universal favor, Daniel Fahrenheit appears to deserve credit for making convenient, portable liquid-in-glass thermometers widely available. Filling thermometers with liquids and sealing them off with a flame must have been a challenging trick to master. Specific information about how this can be done was given by Josef Garriga y Buach and Josef Maria de S. Christobal in their Spanish chemistry text from 1804. The principal steps they give are summarized here as follows:

"*When a thermometer is needed, the first concern is to select a glass tube that is of uniform diameter along all its length. This can be done by filling a small part of the tube with mercury and determining that the length of this sample remains constant as the it moves along the tube.*

One end of the tube is fused in a flame, and when enough is melted, air is blown in from the opposite end to form a small hollow sphere. Next the sphere and a third of the tube are filled with mercury, and the mercury is purged of air and humidity by heating before the tube is sealed off in the flame.

Due to the small diameter of the tube, it is clear that the thermometer cannot be filled by the usual method. Instead, the thermometer bulb is first heated to expand the air inside and drive out much of it. Then the thermometer is dipped open end down into the liquid with which it is to be filled, which will partly occupy the bulb as it cools. Turning the thermometer bulb down again and heating the liquid in the bulb to the boiling point will then drive out more of the air, and then the tube can draw up additional fill liquid as the bulb cools. If bubbles of air remain after the tube is sealed in the flame, spinning the thermometer on a cord will remove them."
(Garriga y Buach, p.65)

(Glass blowing must have been especially challenging in the days when solid fuels such as wood and charcoal were relied on. What was the *lampara de esmaltar* on which our two authors suggest a glass worker could rely?)

At sea level, the reference points of a temperature scale can be determined with melting ice and boiling water and used directly for calibrating thermometers. Rainwater can be collected for this purpose. However, there is an additional problem. Garriga y Buach and De S. Cristobal point out that with different filling fluids, temperatures intermediate to the calibration points may not be in agreement if uniform scale intervals are used.

The way around this difficulty was to eventually adopt a new point of view about the nature of heat and temperature, one suggested by

Sadi Carnot. Along the way, it turned out that a thermometer based on the properties of air like the simple toy described earlier in the chapter was a useful first tool for studying thermodynamics. We will see consequences of this choice in Chapter 7.

The change in electrical resistance with temperature of platinum and other substances can be used to sense temperature. The high melting point and chemical stability of platinum recommend it for high temperature measurements. However, impurities absorbed when platinum is heated to incandescence in an unsuitable environment may affect the accuracy. For lower temperatures (and cost), temperature-sensitive resistors known as thermistors can be used to measure temperature and provide electrical feedback for automatic temperature regulation.

How is temperature regulated in nature? Sunlight is the major source of heat for the earth. Just as a campfire warms campers seated around it, proximity to the sun keeps our world warm compared to planets farther out in the solar system. However, the sun's white-hot surface is a reminder that being too hot can be a problem too. What keeps the earth from overheating?

The earth emits invisible infrared radiation that allows it to maintain a steady range of temperatures across its surface. This radiation is much less intense than sunlight because it is emitted at a much lower temperature, but it balances the input from sunlight by going off into space in all directions and not in just a tiny cone of solid angle half a degree across the way sunlight comes to us.

Sunlight is dazzling, but how do we learn about invisible heat radiation? For starters, our skin allows us to sense heat coming from an open fire. This ability is more developed in rattlesnakes and enables them to catch mice in the dark. Infrared security systems suggest what rattlesnakes may sense on their evening hunts. The astronomer William Herschel used thermometers to detect infrared radiation in the form of a component of sunlight that could be observed beyond the red end of the sun's visible spectrum that Isaac Newton had studied. (White, 2012)

The temperature distribution needed for heat balance between incoming sunlight and outgoing infrared radiation is easy to calculate for the surface of the moon, where heated rocks lose heat directly into space. The problem is more complicated on earth because some gases in the atmosphere that transmit visible light to the surface are less transparent at wavelengths where thermal emission is strongest. Instead of immediately being returned to space, energy from sunlight that is re-emitted by heated structures, rock, and soil goes into heating the atmosphere close to the ground. Because of infrared absorption, the lowest level of the atmosphere becomes a sort of blanket. (Youmans p. 53) However, this blanket has no threads holding it together, so heat can move up into the sky by what is called convective transfer, until at last the air becomes thin and transparent enough for radiative transfer to complete the sequence of absorption and re-emission by radiating energy far away into the depths of space.

The lower atmospheric zone of convective transfer is where much of our weather happens, and it is susceptible to mathematical modeling: For example, a hurricane can be compared to a giant, low pressure gas turbine running off the latent heat of condensing water vapor. (Emanuel, 2006)

The way infrared radiation transfers heat can seem puzzling and counterintuitive. For homeowners in the Southwest, there is no good paint available to reflect the heat from an approaching brush fire because available forms of visibly white paint absorb the intense infrared radiation from a fire.

With sunlight, the situation is more straightforward. White paint on a roof helps control solar heating, especially when there is no outside air circulating in the attic space between roof and ceiling. Roofs of yellow school busses are sometimes painted white, apparently for this reason. It is possible to achieve the opposite effect and locally increase solar heating. Focusing the sun with a magnifying glass works even when the substance being heated is in a glass container. The chemist Lavoisier did this to study chemical reactions under controlled conditions. The lens makes the sun appear much larger from the point of view of whatever is being heated, so the effect is somewhat like taking the sample closer to the sun in a spaceship. Systems of light-weight mirrors can be used in place of lenses to concentrate sunlight for generating electric power on a commercial scale.

To heat a flat panel, one can paint it black, insulate its back side, and mount it in sunlight behind one or more sheets of glass to limit convective cooling. Kilowatts of heating are potentially available from a single rooftop, but installations can be awkward. This approach may disappoint if it is applied to warming a swimming pool, for it makes cleaning more difficult.

Chapter 3
Heat and How It Gets Around

The theory of heat became more quantitative in the eighteenth century after the Scottish chemist Joseph Black pointed out the distinction between heat and temperature. He did this by noting that heat must continue to flow as ice melts even though the temperature of ice and meltwater remains close to freezing throughout the process. He called this latent heat as opposed to sensible heat, which produces a temperature increase that can be measured with a thermometer. Black noted that, without the phenomenon of latent heat, terrible floods would result if every bit of snow and ice were to melt the moment the temperature rose to freezing. (Williams, 2012)

From one point of view, heat is what we use to control temperature. If we are cold, we seek a source of warmth, like a fire, and we have ways of preventing heat from slipping away, which we can do by using a blanket, for example. Ice can be used to draw heat away and produce a lower temperature. Yet both the chemist Black and later the physicist Sadi Carnot were intrigued by situations where heat was clearly flowing while temperature change was minimal.

Heat flows naturally and spontaneously from hot to cold, so an insulated object settles to a uniform temperature, and objects originally at different temperatures settle to a common intermediate temperature when surrounded by insulation. Sometimes one member of such a pair will be so much larger that its temperature appears to remain relatively unaffected by heat exchange with the smaller. However, in the simplest approach to quantifying the flow of heat, two objects or samples are required that start from different temperatures and settle to a common intermediate temperature when a barrier to heat transmission is removed. Independent of the type of thermometer selected, one can conclude that heat no longer flows once equilibrium is established.

This is the situation described by Josef Garriga y Buach and Josef Maria De S. Christobal in their 1804 introduction to calorimetry where equal quantities of water at different temperatures are mixed and then seen to settle at the average of the two temperatures. In this case, an accurate prediction can be made with the help of algebra when the masses as well as the initial temperatures are unequal. Graduated containers to measure volume, balances to determine mass, and calibrated thermometers are the apparatus required. (Garriga y Buach, p. 74)

So the amount of heat that flows can be measured by how much it raises the temperature of a known volume of water. It is an easy method to employ with familiar solids and liquids that can be kept in open containers at ambient pressure and temperature. The utility of thermal insulation makes it plausible to assume that heat is conserved as it spontaneously leaves one body and enters another, and one can rely on the well-known definition that one calorie of heat raises the temperature of a cubic centimeter of water one degree centigrade. (Sears, p. 271) Heat does not need to be recognized as a form of energy for this approach.

Traditionally, the heat capacity of a substance is the amount of heat it takes to raise the temperature of a unit mass by one degree of temperature. Prior to the introduction of an energy equivalent for heat, heat capacities had to be given by their ratios to the heat capacity of a reference substance, usually water.

In order to measure the heat capacity of a substance, a sample of known mass can be heated to a known temperature and placed in a volume of water of equal mass at ambient temperature. It is convenient to do this in a covered plastic foam cup, which provides the needed outer layer of insulation. The heat exchange between reference and sample is assumed to be equal and opposite, so the heat capacity C of the sample is found by solving

$$C(sample) * deltaT(sample) = - C(water) * deltaT(water).$$

provided the masses are equal.

It is assumed that the total amount of heat remains constant when heat passes by conduction from one body to another. The assumption that heat is exchanged in equal and opposite amounts as the water and sample approach equilibrium is crucial for this determination. Notice that the measurement should be repeatable, but the process cannot be reversed since heat flows from hot to cold.

Consistent ratios of heat capacities can be obtained if the water is eliminated from the measurement process and flat samples of different solids initially at different temperatures are placed in direct contact between insulating layers to compare their heat capacities.

Measuring heat capacity for gases is more complicated because they are so much less dense than liquids and solids at ambient pressure. Also, it makes a significant difference whether they are confined at constant pressure or constant volume. In Chapter 7 we will return and take a closer look at the thermal properties of gases.

It is significant that a transient phenomenon--the flow of heat-- produces a change in a property of a sample--its temperature--that can persist in time. Our ability to apply heat in reproducible amounts and produce temperature change more or less at will gives us part of the control we need to "practice" thermodynamics. Converting work to heat by friction and producing work by the expansion of a gas sample are other essential experimental methods.

Approximate values for relative heat capacities at near ambient temperature are shown in the following table where water serves as a reference for the other materials. The larger the heat capacity, the longer a sample of a given mass must be left on a stove to produce a given temperature rise. Water would need to be heated for about ten times as long as an equal mass of iron (for example), because the water would absorb about ten times as much heat for a given rise in temperature. (Sears, p. 274):

Relative Heat Capacity

Substance, Heat capacity for equal masses relative to water:

Water, 1
Ice, 0.55
Aluminum, 0.217
Iron, 0.113
Copper, 0.093
Silver, 0.056
Mercury, 0.033
Lead, 0.031
Glass, 0.20

(Unfortunately, data display in columns is not supported by this ebook format.)

In the days of Joseph Black and Lavoisier, this seemed about all there was to say. Only in the nineteenth century did James Prescott Joule measure and find that a calorie is about 4.186 N m or 4.186 joule. He did this by using a falling weight--the iconic source of mechanical energy--to stir water and by measuring the temperature rise that resulted. (Sears, p. 272) But we are not quite ready to go there yet. The table just presented provides useful information even if we don't recognize heat as a form of energy and try to think of it as a separate, mysterious substance, which is what many did at the close of the eighteenth century.

Lavoisier was implicated in tax farming by the Ancien Regime, and his life was swept away in the French Revolution prior to the work of Benjamin Thompson, which will be described in the following chapter, where heat will be treated as a form of energy. The inability to distinguish between the result of heating by friction and heating by conduction eventually led to the first law of thermodynamics.

Chapter 4
The Mechanical Equivalent of Heat

Redundant calorimetry measurements of the kind described in the last chapter give consistent values for the relative heat capacities of different substances, and these experiments strongly support the idea that heat, whatever it is, can be conserved as it passes down temperature gradients from one substance to another. Until well into the nineteenth century, many scientists thought heat might be a separate substance in its own right that might be referred to as caloric, but a problem with this hypothesis appeared when it was realized that friction produces apparently unlimited quantities of heat for as long as mechanical work is supplied--where could it all come from?

In around 1798, American expatriate Benjamin Thompson was in Munich witnessing the manufacture of cannon possibly destined for fighting in the long aftermath of the French Revolution, which had already taken Lavoisier's life. Thompson had acquired the title of Count Rumford, by which he is often known. This suggests he had status sufficient to account for his presence in an armaments factory.

At a Munich arsenal, Thompson watched dull boring tools raise temperature while failing to cut metal, and he realized there was apparently no end to the heat patiently plodding horses could generate by steadily rubbing blunted tools against unyielding metal. Thompson found that the temperature near the tool-work interface could be raised to the boiling point of water and maintained there. Where was the heat flowing from? The unlimited quantity that could be generated in this way with no apparent source suggested that heat is our perception of motion on a microscopic scale and perhaps more akin to the turbulence in flowing water rather than being a distinct substance in its own right. However, it seems unlikely that Thompson was able to think through all the possibilities we associate with his observations today, and his contemporaries were not immediately persuaded to adopt a new point of view. (Garriga y Buach, p. 45)

In Thompson's experiment, heat was produced as a blunt boring tool scraped inside a metal cannon barrel. One can picture this with the cannon mounted in different orientations and with various arrangements for advancing the boring tool, which could be powered by a horse walking in a circle and turning something like a capstan adapted to this purpose. (Was the cannon barrel horizontal or vertical? Was the bore precast, or were the blanks solid metal? How was the drill head made to follow a straight line as it enlarged the bore? These are practical questions that are seldom addressed in discussions of Thompson's work. The similar problem of boring steam engine cylinders is discussed briefly in Chapter 9.)

The rotating drill bit was a source of heat like a red-hot cannon ball placed in the barrel. The difference is that a mass of metal would quickly cool, and it would soon need to be changed out for another one fresh from a furnace to keep water boiling. Of course, some heat traveled along the shaft of the boring tool as well as through the metal casting.

Thompson's work suggested that mechanical work and heat might be different forms of energy. In his novel and eventually fruitful interpretation, work was being transformed into heat at the boring interface. James Prescott Joule put Thompson's idea to the test by measuring whether the amount of work required to produce a given quantity of heat (that is, to produce the same temperature rise in a sample) remained the same from one experimental test to the next. He was able to show that a given amount of mechanical work converted to heat will consistently raise the temperature of a sample of water by the same amount each time it is applied. The traditional way of producing work he used was to allow a known weight to drop a known distance while pulling a cord wound on a drive wheel. The mechanical work was converted to heat by turning a paddle wheel immersed in a known amount of water in an insulated container. (Sears, p. 270)

It's a good idea to review simple physics and consider how the work done by a descending weight is defined and calculated. We start by considering what happens when an object is allowed to fall freely. While air resistance remains small, its downward speed increases at about 10 meters per second per second. Therefore, in accordance with Newton's law that force is equal to mass times acceleration, it follows that a gravitational force of about ten mks units must act on a kilogram mass. The mks force unit is called the newton (abbreviated N), and we can correctly say that a kilogram mass weighs ten newtons unless greater precision is required. Mechanical work is defined and calculated as force times the distance through which it acts, so a kilogram mass descending a distance of one meter at a steady speed will do approximately 10 newton meters or joules (abbreviated J) of work. (So 10 J is roughly the work required to lift a liter of milk from floor to counter top.) This energy could be applied to heating water by means of a paddle wheel apparatus Joule devised. He found the mechanical equivalent of heat for water to be about 4180 J kg^{-1} K^{-1}; that is, 4180 joules of work raises the temperature of a kilogram of water by one kelvin or degree centigrade. Joule may have worked with a different energy unit, but the newton meter or joule (J) was named in his honor.

Now it is possible to add another column to our table of heat capacities for substances near ambient temperature, one where heat capacity is given as an energy per unit sample mass per unit temperature change. As before, C is used to represent heat capacity in units of joules per kilogram per kelvin temperature unit:

Heat Capacity

Substance, Relative heat capacity for equal masses, C(J kg^{-1} K^{-1}):

Water, 1, 4180
Ice, 0.55, 2310
Aluminum, 0.217, 908
Iron, 0.113, 473
Copper, 0.093, 389
Silver, 0.056, 234

Mercury, 0.033, 138
Lead, 0.031, 130
Glass, 0.20, 840

The quantity name--heat capacity--suggests that the ability to absorb heat might be advantageous. This would be the case if one were using rocks heated in a fire to boil water in a container that could not be placed directly over the flames, a practical method that today seems more like a recourse of desperation.

The values in the third column of the table are derived from those in the second using 4180 J kg^{-1} K^{-1} as the value of the mechanical equivalent of heat that Joule measured for water at room temperature. Joule's experiment is difficult to carry out because of the relatively large amount of mechanical work needed to produce a detectable temperature change. Also, if air circulates in the apparatus, evaporative cooling may mask the effect. For a laboratory exercise today, it may be easier to measure the mechanical equivalent of heat by using dry friction to heat a rotating aluminum or copper cylinder.

If force times distance defines work, then work divided by time defines power, which has the newton meter per second (N m s^{-1}) or watt (abbreviated W) as its mks unit.

How did it originally come to be realized that a force multiplied by the distance through which it acts gives a useful measure of work? Part of the answer is that whatever efficient simple machine--a block and tackle, an inclined plane, a bicycle--is used to help accomplish a task, the power available from a man or a beast of burden remains about the same when measured in this way. Walking on a smooth surface is an efficient form of locomotion, and one may stroll along briskly on level ground at around one meter per second well enough, but at a slope of one in ten, the effort becomes noticeable. A 70 kg person weighs about 700 newtons N, so the force of the person's weight resolved along a 10% grade that must be overcome is 70 N. At a speed of one m/s, the rate at which work must be done to propel a hiker up the hill is 70 N m s^{-1} or 70 W, which requires almost a

100% increase over the human resting metabolic rate given in references as around 87 W. (Harrison) If a typical hiker attempts to go much faster up a 10% hill, they will become out of breath and lose some of the lead they gained when they stop to rest. In the late 1700s, James Watt applied this sort of reasoning to work done by horses when he needed a measure of power for marketing the improved steam engines he was manufacturing in Manchester with Matthew Boulton. The value he chose converts to 746 W per horse, perhaps ten times as much as one might expect from a convict on a treadmill. (Lienhard)

The reproducibility of Joule's measurements of the mechanical equivalent of heat helped complete Thompson's work and supports the ideas that heat is a form of energy and that energy is conserved. The first of these seems an easy call today, but the proposition that energy is conserved and only transformed from one form into another wasn't accepted all at once. Later on, when we encounter entropy, it may seem as non-intuitive a concept as energy might have two centuries ago, even though energy and entropy stand experimentally on equal footing as valid descriptors of the natural world.

Another nineteenth century physicist, Dr. J. R. Mayer, independently proposed the equivalence of multiple forms of energy. He studied the behavior of air and noted the difference between the coefficient of heat at constant pressure and the coefficient of heat at constant volume. Their ratio is constant independent of pressure, which suggests a relationship between the internal energy of a gas and its work of expansion. (Sandfort, p. 76) We will encounter the distinction between the coefficients of heat at constant pressure and constant volume when we analyze the Clement-Deshormes experiment in Chapter 7.

Chapter 5
The First Law of Thermodynamics

In the last two chapters, we have seen examples of irreversible changes. In calorimetry, heat flows spontaneously from hot to cold, and in Thompson's cannon boring experiment, work is converted to heat at the place where rubbing occurs and never the other way around. Richard Feynman discusses reversibility and irreversibility in his Lectures on Physics. (Feynman, Ch. 46) In purely mechanical problems, he notes, processes are often reversible in time. But this is not generally the case in thermodynamics. To understand the reason, we must consider phenomena on an atomic level. When we do, Feynman concludes that "It is the change from an ordered arrangement to a disordered arrangement which is the source of this irreversibility." Large scale phenomena can exhibit irreversible behavior, too. For example, turbulence in the wake of a sailboat or glider cannot be undone by reversing the direction of travel.

The way heat refuses to flow "uphill" from cold to hot is a major manifestation of the microscopic irreversibility discussed by Feynman.

Brownian movement seen through a powerful microscope as the jiggling of tiny particles suspended in water is primary evidence for random behavior on a microscopic scale. Historical arguments for the existence of atoms and molecules are based on Brownian movement. (Bernstein, 2006) Later, in Chapter 14, we will apply the principle of naturally increasing disorder to thermodynamics using an ideal gas as our model and following the example of Ludwig Boltzmann.

Understanding of the nature of heat can be sought in introductory physics and chemistry textbooks at the university level. A compact discussion on an advanced level is given by E. U. Condon. (Condon, p. 5-3) Here, we will attempt to cover some of the same ground without appealing to such advanced mathematics.

Heat is associated with the spontaneous flow of energy from hot to cold in ordinary thermal conduction. Sample heat Q increases when heat enters the sample spontaneously by conduction. Friction generates heat that spreads by conduction from where it occurs. Radiative heat transfer fits this model also, but the details are modified because direct physical contact between the sample and a source or sink of heat is not required.

The first law of thermodynamics establishes a link between large scale phenomena and the microscopic motion of individual atoms. Gas expanding against a piston in a cylinder can convert heat into work. If no heat flows from hot to cold during the process, then the work is assumed to come at the expense of the random thermal energy of the atoms of which the gas sample is composed.

Sample heat can sometimes be measured by observing the temperature change of a sample or its surround, and work is measured by the product of a force times the distance through which it acts. To keep track of all the energy, the internal energy of a sample of substance must be introduced, even though it is not observed directly. The first law of thermodynamics describes how these quantities are related:

Sample heating is equal to the change in sample internal energy plus the mechanical work the sample does by expanding.

This form of the first law describes thermal effects where gas or liquid samples perform work on pistons moving in well-fitted cylinders. We will often imagine work as coming in P times dV increments, where P is pressure and dV is a small relative change in volume. But work can be done in other ways that we will need to consider. Heat and internal energy can be transformed into the kinetic energy of a fluid emerging from a nozzle, and this kinetic energy can be harnessed by spinning turbine blades. Heat can manifest itself virtually by a phase change as when ice melts or water boils.

Robert G. Fovell in his Great Courses lectures on meteorology suggests that heat can often be thought of as a transient flow of energy, whereas variables such as pressure, volume, and temperature are quantities that can persist indefinitely if a sample of material is left undisturbed. Similarly, work represents a transient movement against resistance. Internal energy, on the other hand, falls in the same category with pressure, volume, and temperature.

Benjamin Thompson's cannon boring experiment showed that mechanical work converted to heat by friction increased temperature just as if it came from a cannon ball that had been heated in a furnace and placed in the barrel. Friction heating, unlike mechanical work, is not time reversible and always results in a flow of heat away from a work face, typically in two directions simultaneously. Friction contributes as a way to supply heat, but the inverse process, where heat is spontaneously converted to work on a large scale, is completely impossible.

Radiative heating is akin to contact heating because net radiative transfer is from hot to cold even when complex optical systems are used to control the overall effect, as in solar power stations. (Radio transmitters and lasers are exceptional because these sources operate far from thermodynamic equilibrium.)

Here our mathematical statement of the first law of thermodynamics emphasizes the importance of observing small relative changes in thermodynamic quantities and calculating larger changes by numerical integration. We use d instead of the traditional capital Greek letter delta to write:

$$dQ = dU + dW,$$

where Q, W, and U stand for heat, work, and internal energy respectively. If the total change in the heat (for example) is specified, then the incremental changes dQ, dW, and dU can be treated as so-called finite differences that are reduced toward zero in the course of a calculation as more and more are summed to produce estimates of increasing precision. When the first law is used in this incremental form, computers provide insights previously reserved for students of calculus, and the presentation in this book is organized differently from the way it would have been in a thermodynamics textbook from half a century ago when students seldom had access to digital computers.

Like the quality of mercy, the first law applies to both source and sink in a spontaneous heat exchange. The quantities of heat are equal and opposite, and the divisions into work and internal energy suit the source and sink respectively.

For a so-called ideal gas like air, the internal energy U can be thought of as the sum of the kinetic energies of the molecules, which move about independently, and this internal energy is proportional to the Kelvin temperature. The performance of heat engines that use air as the working fluid is relatively easy to describe and predict mathematically.

Unlike heat and work, internal energy seems to evade direct observation, although beams of atoms can be produced and their velocities measured using high vacuum technology. (Born, p. 63)

Often it will be convenient to write the first law as

$$dQ = dU + P\, dV$$

Words can express the sign convention used here: "Heat entering a sample equals the sum of the increase in its internal energy and the mechanical work it performs by expanding against a piston in a cylinder." (The change in volume V takes into account both the area of the piston and the distance it moves.) In the absence of a piston and cylinder, expansive work can be done against atmospheric pressure, and its effect is to raise the height of the atmosphere a negligible amount.

Note that the the internal energy U does not include the collective kinetic energy of a sample. This collective kinetic energy becomes significant when liquid or gas flows through a nozzle. Later we will consider cases where heat flow is related to changes in phase due to melting or evaporation.

After Benjamin Thompson's work, the caloric theory hung on until well into the nineteenth century. Change came at last with William Thomson's and Rudolph Clausius' analyses of heat engines working in thermodynamic cycles. Thomson, later Lord Kelvin of temperature scale fame, concluded that the work produced by the gas in the engine cylinder during a complete cycle was equal to the difference between the amount of heat it absorbed from a hot reservoir and the amount it discharged to a cold reservoir. (Sandfort, p. 87) Let us look at this result from a slightly different point of view by writing the last equation as

$$dQ - P\,dV = dU.$$

If the terms on the left are summed around a closed path in thermodynamic variable space (that is, around a complete thermodynamic cycle of an engine, for example), then the result approaches zero in the limit as the increments are made smaller. The sum of the changes of the internal energy U around a closed path is always zero, making internal energy an intrinsic property of the substance independent of its history. Heat and work, on the other hand, are manifestations of ongoing change: If either heat or work is plotted on the same diagram in place of internal energy, the curve will not retrace itself but move upward or downward in a repeating pattern as the engine continues to operate.

This result establishes internal energy as an intrinsic property of a gas sample along with temperature, pressure, and volume. We will follow these ideas further in Chapter 7.

The kinetic theory of heat took a long time to gain acceptance. One can at last find heat described as a mode of motion in a chemistry textbook from 1875. Science was at another major turning point by this time, for elsewhere in the same text, one finds the attribution of dangerous infections to microscopic organisms. (Youmans, p. 63 and 302)

At the turn of the nineteenth century, people could not speak so confidently of the kinetic energy of a gas molecule as we do today. As Garriga y Buach and De S. Christobal pointed out, measurements of heat and temperature were more subjective than measurements of time and distance, while the concept of kinetic energy as a generally useful quantity was still emerging. However, starting in Chapter 7, we will in effect assume that "true" temperature T is proportional to the average kinetic energy of a molecule in a sample of gas. Unless we do, later statements about a new state variable called entropy will be difficult to justify.

Chapter 6
Heat Transfer

We have established that spontaneous thermal energy transfer down temperature gradients is one of the signature properties of heat, but have said little about the process, since it happens only when equilibrium is not established, and much of thermodynamics is the study of equilibrium states. When equilibrium cannot apply, it is often convenient to analyze steady state heat transfer where the rate of energy transfer remains constant and the temperature distribution does not change with time.

The effectiveness of heat engines and refrigeration systems depends on transferring large amounts of heat with a minimal temperature drop across interfaces that often must resist deformation by pressure. If we think of a non-steady condition with a temperature gradient shifting gradually with time after heat begins to flow, then we would say heat travels very slowly compared to sound. This can be observed by feeling a ceramic mug warm up after being filled with hot coffee. The change would happen more quickly with an aluminum cup in place of a ceramic one. What are other ways to speed up heat transfer when it is needed?

In addition to enlarging the area where heat exchange is occurring, a way to increase the rate of heat transfer is to increase the temperature difference across the barrier keeping the fluids separated. However, we won't estimate the sizes of the heat exchangers required for the refrigeration systems we discuss and will only reflect on the significant amount of energy needed to circulate fluids through them. But minimal attention here does not mean heat exchangers are boring. For example, steam boiler technology was pushed to its limits as internal combustion was introduced and eventually proved essential for the development of aircraft.

Calorimetry was introduced in Chapter 3. The method of measurement described there relies on the way objects originally at different temperatures come into equilibrium at a common temperature when placed together in an insulated enclosure such as a foam plastic cup. The time it takes for this process may be increased or diminished by the choice of the material separating the substances.

We may need to ask how quickly heat energy will move through a layer of material if a steady temperature difference is maintained from one side to another. Heat transfer through the cylinder walls in a water-cooled internal combustion engine is a good example. If the walls are too thin, the engine will be difficult to manufacture and perhaps prone to failure. If the walls are too thick, the power-to-weight ratio will suffer and more water may need to be circulated through the cooling passages to maintain the optimum temperature. The rate at which heat passes through the metal and into the engine coolant is proportional to the temperature difference and the thermal conductivity and inversely proportional to thickness, and the same applies for other applications. The following table gives the approximate thermal conductivity in Joule meters per square meter per second per kelvin temperature difference for a number of materials:

Thermal Conductivity

Material, Conductivity (J m m^{-2} s^{-1} K^{-1}):

Water, 0.6
Aluminum, 205
Copper, 385
Steel, 50
Glass, 0.84
Brick, 0.63
Concrete, 0.84
Wood, .084
Cork, .042
Air, .024

Note that for air and water, temperature must increase upward to avoid convection. (Sears, p. 288)

Here thermal conductivity might be given as watts per kelvin per meter, but in the context of this chapter especially, it is advantageous not to lump units for distinct quantities such as area and thickness. The jumble of symbols tells us that the amount of thermal energy in joules that passes through a sample in the form of a slab is equal to the conductivity constant in the table times the area in square meters, times the temperature difference through the slab in kelvins, times the time in seconds, and at last divided by the thickness of the sample in meters. If one carries the units along and keeps track during a calculation of the total heat energy transmitted, unit symbols cancel out one by one as the calculation proceeds until only the joule will remain as the unit for the final result giving the total amount of energy.

It is not difficult to adapt the previous table to a change in units. If, for example, time were given in hours instead of seconds, one would multiply the conductivity in the table by the measured time multiplied by 3600 to get the correct result for the amount of heat in joules transferred through the slab. The table can be re-written in advance to take this change into account: For each substance, the value in the table would need to be multiplied by 3600 to account for the change from measuring time in seconds to measuring it in hours.

Thus a general technique for dealing with unit changes is to substitute for the symbol of the unit being changed using the value of the first unit in terms of the new unit and then perform the indicated multiplication or division. For example,

50 km/hour = 50 (1000 m)/(3600 s) = 50 * 1000/3600 m/s = 13.9 m/s.

Systematically making substitutions in this way allows researchers working with different unit systems to check and verify the correctness of one another's results. The units needed to describe different physical quantities such as velocity and power are represented by combinations of the fundamental units for mass, length, and time according to the arithmetical operations employed. The process of unit conversion takes care and concentration, so we are trying to detour around it in this ebook by sticking to mks units as much as possible.

The remarkable Joseph Fourier is credited by James Clerk Maxwell for pointing out the importance of using units consistently in engineering calculations. This occurred roughly a century after Newton had shown in a remarkable synthesis how numbers representing distances, velocities and accelerations could be combined mathematically to describe and predict the motion of planets. (Maxwell, p. 1)

Each term in a mathematical equation expressing a physical law such as force equals mass times acceleration must have a unit equivalent to those of the other terms. (Terms are separated by +, -, or = signs as in the first law of thermodynamics, where heat, work, and internal energy all have the same unit.) Otherwise, performing a unit conversion could destroy the numerical equality that is supposed to hold independent of the system of units used by an observer.

This same technique is extended to what seem like ambiguous cases where there is no physical reference unit, as when one is converting from one temperature scale to another or from degrees to radians for measuring angles.

The root of the mean square error (*RMSE*) estimate of uncertainty is an example of Fourier's insight at work. The formula is

$$RMSE = \text{Sqrt}(Avg[(o-a)^2]),$$

where the *o* are the observed values and *a* is the average of the observed values, so the root of the means square error is always greater than zero. Taking the square root in this way assures that the error estimate will have the same physical dimensions (or unit) as the quantity it applies to. This makes the ratio of the root of the mean square error to the average value independent of the choice of unit used for the measurement. (Feynman, Ch. 6)

The range of thermal conductivities shown in the last table is noteworthy. High conductivity is useful in cookware and in the heat exchangers that we will need to introduce later on, while poorly conductive materials are needed for thermal insulation. The insulating nature of rock helps to shape the world. It is believed volcanoes can erupt molten rock because thermal conduction does not keep up with heating due to radioactive decay deep in the earth's interior. This same limitation renders underground storage of large quantities of highly radioactive waste problematic.

An introduction to transport phenomena is given by G. K. Batchelor in his text on fluid dynamics. (Batchelor, p. 28)

The conductivity of a poorly conducting material like cardboard can be measured by placing a sheet between slabs of highly conductive metals like copper or aluminum initially at different temperatures. The "sandwich" is wrapped in an insulating blanket and the slab temperatures are tracked. If the temperature difference is found to decay smoothly with time while the metal slabs remain at relatively uniform temperatures, then the parameters of the system can be used to calculate a value for the thermal conductivity of the intervening layer. We can refer to this situation as (near) steady state heat exchange, and this model can be used to describe what happens in heat exchangers when the conducting layers are formed into tubes to prevent fluids at different temperatures from mixing. Turbulent flow is desirable to increase the rate of heat exchange, but one ought not assume that the full temperature difference acts across a solid layer separating the media.

Often the size of a heat exchanger represents a compromise. Suppose air circulating through a house is cooled from an indoor ambient temperature of 298 K to 286 K by the evaporative heat exchanger in the plenum of the duct system that distributes cooled air to the cooling vents in the different rooms of the house. To achieve a compact system, it may be expedient to cool the refrigerant circulating in the tubes of the heat exchanger close to freezing, well below the temperature at which air will emerge into the house. This is an inevitable loss that is accepted as part of what is needed to produce an optimum system.

Counter-flow heat exchangers are an important exception to the situation described in the last paragraph. In this case, tubes are placed in close thermal contact along their entire lengths and the initially hot and cold fluids flow in opposite directions. (Copper tubes might be soldered together along their entire length and enclosed in insulation.) If the process is properly scaled and executed, the fluids emerging will have switched temperatures with relatively good precision, so this effect is almost reversible, unlike the calorimeter measurement described in Chapter 3, where the samples begin at different temperatures and come into equilibrium at an intermediate final temperature. The counterflow principle helps explain why one can melt steel in an open-hearth furnace but not a steel paperclip in the gas flame of a kitchen stove. Some animals benefit from the counterflow principle, and it is an aid in reaching the cryogenic temperatures required to liquify air. An interesting demonstration could be set up in school laboratories equipped with hot and cold running water.

Here we might apply a modification of the counterflow principle to calculate the heat cost for direct cooling. For example, if warm air is slowly blown through a long tunnel filled with buckets of ice closely spaced on open racks, air at near freezing temperature will continue to emerge at the far end until all the ice is melted.

Chapter 7
Calculating with Ideal Gases

Thermodynamics provides a quantitative description of homogeneous matter in situations where variables including temperature can be observed or controlled. Often two parameters can be related to the temperature by means of a so-called equation of state that describes a sample of substance in equilibrium so time is not an explicit variable. Thermodynamic equilibrium requires that temperature be uniform in time and space; nevertheless, thermodynamics can be applied to systems that change with time as they pass through a sequence of near-equilibrium states. (Batchelor, p. 20)

Ordinary atmospheric air is a good example for introducing the equation of state concept, and pressure P, volume V, and Kelvin temperature T are familiar choices for the thermodynamic variables. Bubbles rising in water suggest that a gas sample can have a definite volume. The dependence of gas volume on pressure is illustrated by the Cartesian diver, a toy diving bell that just floats in a jar completely filled with water and sinks when a flexible, tightly fitting cap is depressed. The way gas pressure changes with temperature at near constant volume is the basis for the simple thermometer described in Chapter 2. Later in the book, state variables (or state functions) other than pressure and volume will serve as principal thermodynamic coordinates.

Three interrelated variables are sufficient to describe the thermal behavior of familiar gases. Later, we will see how this approach can be extended when there are two coexisting phases of the same substance in contact such as ice and water, water and steam, or a liquid refrigerant and its vapor.

The important thermodynamic quantities heat Q and work W are not state variables, and they are agents of change as a thermodynamic system transitions from one equilibrium state to another.

The molar relationship $PV = nRT$ is the familiar equation of state that is often used for air and the well-known gases hydrogen, helium, nitrogen, and oxygen that are relatively difficult to liquify (compared to chlorine and carbon dioxide, for example). Here n is the number of moles of gas. Later we will consider the behavior of other gases that are close to the point of condensation and require a different treatment--steam coming from a teakettle is an example.

In the absence of an acceptable equation of state, temperature contours can be shown on a pressure versus volume chart the way landscape elevations are shown on topographical maps, or the information can be presented by entries in a table. Even though the ideal gas law $PV = nRT$ is a starting point for what we are doing here, we will not always rely on the existence of explicit equations of state to underpin explanations.

In contrast, we will often model thermodynamic processes by starting from a known condition and following through with numerical integration to find the limiting sums of incremental (or indefinitely diminishing) changes. In many situations where the total change of heat (for example) is bounded, accuracy can be achieved by making the changes smaller and summing more of them. This relatively elementary approach relies on the availability of digital computers and is suited to the incremental form of the first law of thermodynamics from Chapter 5,

$$dQ = dU + dW$$

Depending on the circumstances, changes such as *dQ*, *dW*, and *dU* can be specified arbitrarily, measured, or calculated. They may be thought of as small increments to the respective quantities, and sometimes they will represent so-called differential changes in specific mathematical functions. In this ebook, lower case d must stand in for upper case Greek delta, which is frequently used to represent an increment in a quantity.

As presented in calculus texts, a differential change *dx* is understood as a small change in the variable *x*. If another variable *y* depends on *x*, then the component of the change in *y* that is proportional to the change *dx* in the limit as *dx* diminishes is called the differential *dy(dx)*, where *dy(dx)* can be read as "d y of d x." For example, if $y = x^2$, then *dy(dx)* = 2 *x dx*, while the total change includes what is called a higher order term in *dx*. (Hardy, 1960) In this presentation, we do not attempt to use differential and integral calculus to eliminate explicit reference to incremental quantities from our analysis. Reliance on numerical integration with the help of a computer underlies the presentation.

In thermodynamics, a variable is a state function if the sum of its incremental changes around an arbitrary closed path in thermodynamic variable space computes and measures close to zero.

In this case, it may be possible to represent it as an explicit function of other state variables.

Of the first law quantities Q, W, and U, only the internal energy U is a state function. Let us consider an important consequence. Suppose dW is in the form of expansive work and given by

$$dW = P\, dV$$

Then, for a closed path in the PV plane,

$$\text{Sum } dQ = \text{Sum } P\, dV,$$

since Sum $dU = 0$ for a closed path.

This is the basis of the indicator diagram method used to analyze the performance of engines that work in cycles: If the axes represent pressure and volume, then the area of the diagram is proportional to the work performed by an engine in a single cycle, which can be equated to the net heat absorbed by a gas sample doing the work. (Sears, p. 303, and Ripper, Ch. V)

It is easy to imagine a fixed sample of gas following a cycle of expansion and contraction in a cylinder fitted with a moveable piston as heat enters and leaves by conduction. For air conditioning, we can also apply the cycle idea to standard quantities of refrigerant passing one after another through pipes from compressor to condenser and so around a cycle, even though intermixing must inevitably occur between samples flowing sequentially. However, a closed indicator diagram does not establish that a heat engine is precisely reversible in the thermodynamic sense. This is because the direction of thermal gradients in heat exchangers must be reversed when a heat engine is operated as a refrigerator.

Adiabatic expansion

A volume change without heat flow is an important thermodynamic process for us to analyze because it can produce cooling. Adiabatic expansion of air in a cylinder fitted with a piston is produced by drawing out the piston too quickly for a significant amount of heat to flow from the cylinder wall and influence the result. For a mole of air,

$$PV = RT,$$

and we can introduce the expression for the molar internal energy of air,

$$U = 5/2\, RT,$$

to use in conjunction with the first law of thermodynamics. Notice that the internal energy of air depends only on temperature--this will help us here and cause a difficulty later on. The pioneers of thermodynamics were unaware of the ideal monatomic gases helium and argon, which were not successfully isolated and identified until toward the end of the nineteenth century. For thermodynamics, argon and helium are notable because their molar internal energies are given more nearly by $3/2\, RT$ rather than by $5/2\, RT$ as in the case of the more familiar ideal gases hydrogen, nitrogen, and oxygen.

The first law is written in incremental form, first (as in Chapter 5) as

$$dQ = dW + dU$$

and then more specifically as

$$dQ = P\, dV + 5/2\, R\, dT$$

because an increment of work dW done by the contained gas on the piston can be estimated as the pressure times the area of the piston times the distance the piston moves. Also, we are introducing $5/2\, R\, dT$ as the change in the internal energy of a mole of air corresponding to a change in temperature dT. These results are all we need to predict the temperature changes that can be produced by compressing or expanding a sample of gas. Prehistoric people apparently discovered this effect and used it to start fires prior to the introduction of the fire piston into Europe. (Carnot, p. 16)

When the sum over successive increments is taken, the accuracy can be expected to improve as the increment size is diminished and more terms are summed until other uncertainties make further refinement pointless. If we assume rapid expansion so $dQ = 0$, then $5/2\ R\ dT = -P\ dV$,
and we can introduce numerical integration by calculating the volume change needed to cool air from its ambient condition to the temperature at which water freezes. For this, we sum temperature changes for a sequence of short steps in which the volume is incremented by amounts dV.

In our first numerical integration example, a one mole air sample starts at one atmosphere of pressure or 100000 N m^-2, a volume of 24.8 m^3, and our reference ambient temperature of 298 K. It expands in steps of 0.1 m^3 until the temperature falls below the freezing point of water at 273 K. Here is the program for the calculation as performed with Liberty BASIC software on a Dell Inspiron computer:

```
print "Ad Exp Ch 7 211021"
print "P V T"
P=100000
V=24.8
dV=.1
T=298
R=8310
print P, V, T
```

The volume V is the independent variable, which is incremented repeatedly in steps of $dV = 0.1$ m^3 using the while ... wend command pair as follows:

```
while T>273
V=V+dV
```

At each step, the temperature is diminished by the amount required so the work is done at the expense of internal energy. The resulting change in temperature is proportional to the relatively small change in volume dV:

dT=-2*P*dV/R/5
T=T+dT

Then the equation of state is used to recalculate the pressure, and the process is repeated with the wend command:

P=R*T/V
wend

After the temperature reaches the freezing point at 273 K, iteration ceases, and the the program prints the final values of interest before stopping:

print P, V, T
end

Here is is the entire output obtained from running the program:

Ad Exp Ch 7 211021
P V T
100000 24.8 298
73376 30.9 272.8

(Unfortunately, printing in multiple aligned columns is not supported by this ebook format.)

The calculation shows a relative expansion of 30.9/24.8 = 125% is sufficient to cool air from room temperature to freezing at 273 K. Later on, we will discuss a different approach, the Clement-Deshormes experiment, which is an easier way to observe this effect with simple apparatus. If one attempts the experiment simply by reversing the orientation of a cup leather in a tire pump, both the rapid flow of heat from the cylinder walls and friction heating mask the temperature drop.

The state variable entropy

The important state variable entropy S is determined in a similar way to internal energy; that is, based on observations of changes. (Sandfort, p. 175) When preparing a table of entropy values for a substance, it is necessary to assign an arbitrary reference value for S under agreed standard conditions in the same way 0 C marks the freezing point of water on the centigrade scale. (Masterton, p. 691) Then an incremental entropy change dS can be found by measuring a small amount of heat dQ entering or leaving a sample while observing the Kelvin temperature and then performing the calculation

$$dS = dQ/T$$

There is no convenient entropy meter comparable to a thermometer. For accurate entropy change measurement, only a small amount of heat can be allowed to flow at one time so the change in temperature will be relatively small for each step in the process. By proceeding in this way, it is possible to measure entropy change precisely enough so the changes can be backed out by performing the measurement steps in reverse order. Entropy changes around a closed path in thermodynamic variable space will measure and calculate close to zero when the work is carefully done--otherwise entropy could not be a state variable.

Entropy changes are reversible only from the point of view of an individual sample of matter. This is because the heat must come from outside the sample, so multiple entropy changes are occurring simultaneously. Overall, the net entropy change is never less than zero, and this constitutes one statement of the second law of thermodynamics. We will return to this important topic in the next section of this chapter and again in Chapter 11.

Now let us look more closely at the case where a change in a thermodynamic variable can be found from a specific algebraic expression as well as from the data columns of a thermodynamic table.

Suppose z represents a variable. If its change can be given by

$$dz = f(x)\, dx,$$

it is by default a function of the single variable x. This is the case for the internal energy of a diatomic gas like air where $U = 5/2\, R\, T$.

Suppose instead that

$$dz = f(x)\, dx + g(y)\, dy$$

Here the change in z remains independent of path between two points in the $x\, y$ plane, so z still satisfies the condition for a state variable. This breaks down if we have instead

$$dz = f(x, y)\, dx + g(x, y)\, dy,$$

where the sum of the changes dz around a closed path can no longer be assumed to equal zero.

Let us see where the idea of path independence leads when we calculate dQ for the case of an ideal diatomic gas like air. Then $P\, V = R\, T$, $U = 5/2\, R\, T$, and

$$dQ = dU + P\, dV$$

becomes

$$dQ = 5/2\, R\, dT + P\, dV,$$

which does not assure path independence. However, dividing through by T gives

$$dQ/T = 5/2\, R\, dT/T + P\, dV/T = 5/2\, R\, dT/T + R\, dV/V,$$

which does. (The corresponding formula for a monatomic gas for which $U = 3/2\, R\, T$ appears on the cover of this ebook.) This result is consistent with the idea that summing dQ/T along a path in the pressure-volume plane gives the change in the entropy of an ideal gas. The last equation can be integrated to obtain a formula for the entropy of an ideal gas that includes an undetermined constant term. (Levine, p. 87) Here, however, we shall apply numerical integration for calculating entropy change when we return to this topic in Chapter 11 and see how Rudolph Clausius was able to build on this result and conclude that entropy is a state variable for substances other than ideal gases.

Reversibility and the second law of thermodynamics

Some phenomena are reversible in the sense they can be run backward in time, almost like a motion picture, until a former state is restored. A rock thrown in a vacuum is an example. (Feynman, Ch. 46)

Thermal conduction, on the other hand, is an example of an irreversible process that will run only from hot to cold. In this case, how can entropy be a state variable if each incremental entropy change dS of a sample is calculated from

$$dS = dQ/T\,?$$

To reverse exchanges of small quantities of heat dQ, the temperature of the surroundings of the sample can be changed as needed so that measurements of sample entropy changes can be reversed with good (but never perfect) precision. The temperature uncertainties introduced can be reduced until they seem negligible so far as the entropy measurement of the sample is concerned. The catch is that the sum of all the entropy changes (including those of the reference heat reservoirs) will always be found to obey

$$dStotal = dSsam + dSsur >= 0,$$

which is Rudolph Clausius' form of the second law of thermodynamics. But only the part of the first measurement that directly applies to a selected sample needs to be precisely reversed in a second measurement. (Feynman, Ch. 44)

So, properly measured, entropy appears as a state variable. There are almost reversible processes by which entropy can be thought of as changing gradually in small steps during which heat flows along minimal temperature gradients so that the sample temperature remains almost uniform throughout the process. In this case, the sample moves from an equilibrium state to another nearby equilibrium state. When we measure sample entropy change, we are ignoring the inevitable entropy changes of other heat reservoirs employed in the measurement.

There are also inherently irreversible processes such as flow through a throttle or gas flow into a vacuum through a burst diaphragm. To calculate the entropy change between the equilibrium end points in these cases, it is necessary to trace the entropy change through a series of reversible steps leading from the sample's initial state to its final state. We will consider an important example of an irreversible process in the next section.

Enthalpy

At this point, we also need to introduce the state variable enthalpy H even though the enthalpy of an ideal gas is a physical property that is easy to pass over as uninteresting because forcing air through a throttle does not produce cooling. This poor estimation will be dispelled in Chapters 12 and 13 when we consider enthalpy and entropy together in the context of phase changes and chemical reactions.

The molar enthalpy H is defined by

$$H = PV + U,$$

so it is clearly a function of temperature for an ideal gas. However, for a pure chemical compound consisting of liquid and vapor in equilibrium at a specific pressure, the total enthalpy will depend on how much of the sample has evaporated.

If a slowly moving fluid passes through a thermally insulated throttle so no heat is exchanged and kinetic energy remains negligible, then

$$(P\ V + U)\text{upstream} = (P\ V + U)\text{downstream},$$

and there can be no temperature change with an ideal gas. With a liquid, cooling is possible if a phase change occurs so a mixture of liquid and vapor phases emerges from the throttle. Throttled flow is irreversible and comes with an attendant increase in entropy. (Levine, p. 51)

Trekking the pressure-volume plane

We have introduced a two-dimensional thermodynamic variable space with pressure and volume as the coordinates and described how constant temperature contours can be shown as features on this surface. In this chapter, we are focusing on the thermodynamic properties of a sample of substance in the form of an ideal gas. Contours for other state variables, especially internal energy, entropy, and enthalpy, can also be drawn.

If we are studying an actual gas sample and have the necessary measuring instruments, we can record paths our experiments follow on this surface as the sample absorbs heat, performs work on a piston, or passes through a throttle valve. Some changes we can initiate, and others follow as dictated by the laws of thermodynamics.

By a stretch, we can imagine we are in an airplane leaving a smoke trail, and that the paths of our measurements or calculations laid out on graph paper are the map of adventures in an abstract land of marvels.

The numbers we plot may be either measured values or calculated values based on the laws of thermodynamics, and we are anxious to see agreement between the two approaches. Half a century ago, access to digital computers was restricted, and emphasis was placed on mastery of the advanced mathematical methods used by pioneers like William Thomson, Rudolph Clausius, James Clerk Maxwell, and Josiah Willard Gibbs. The goal was to derive so-called closed form solutions so that graphs could be generated and theories checked by substituting into a mathematical formula. Thus physics students need a strong minor in mathematics. A look into G. H. Hardy's formidable mathematics textbook suggests the level of expectation. (Hardy, 1960)

Here we are committed to solving many of the problems that present themselves by numerical integration, the so-called method of finite differences. This approach has already been demonstrated in this chapter when we calculated how much a sample of air must be expanded to cool it to the freezing point. The examples here are difficult to carry out on a spreadsheet, so computer software resembling the BASIC programming language may need to be downloaded from the Internet.

As we continue to study refrigeration, we will see that it is sometimes helpful to use coordinate pairs other than pressure and volume as the fundamental state variables. Temperature versus entropy and pressure versus enthalpy are important alternate choices.

It is important to hold in the back of the mind that heat and work are distinct from state variables. Heat flows and work is done in order to move about the state variable plane. This is illustrated in the adiabatic expansion calculation earlier in the chapter: In effect, we allowed the gas to expand and do work to obtain the temperature change desired. In this chapter, we will visit this idea again in the section *Examples of calculations with ideal gases*, below.

The simplest refrigeration system we can consider uses air as a working fluid in a configuration that can be operated either as a heat engine or a refrigerator by making a few minor changes. In the next chapter we will study an air cycle air conditioning system and its manifest drawbacks. The practical refrigeration approach that is used instead produces a temperature drop by allowing a liquid refrigerant to pass through a throttle valve at constant enthalpy. Cooling occurs as the refrigerant evaporates, as can easily be demonstrated with acetone, alcohol, or even water. So we must extend our mathematical model to include the simultaneous presence of both a liquid phase and a gas phase. A surprising point to consider is that this type of refrigerator cannot have its cycle reversed so it generates mechanical power as a heat engine.

Following chapters will attempt to make sense of this puzzling clutter of information.

Higher order differences

Because of our reliance on numerical integration of data obtained by measurement or estimated based on mathematical expressions representing physical laws, we are only gradually introducing information that would be presented systematically in an introductory calculus course.

An important idea we rely on is that the increment of the product of two variables x and y is

$$d(x\,y) = x\,dy + y\,dx.$$

When we apply this to the equation of state of an ideal gas

$$P\,V = R\,T$$

we get

$$P\,dV + V\,dP = R\,dT$$

for the result. But what has become of the term $dP * dV$? Wouldn't it be more accurate to write

$$P\,dV + V\,dP + dP\,dV = R\,dT\,?$$

The reason the sum of terms $dP\,dV$ can be dropped from the numerical integration is that it is of so-called higher order than the sums of the other terms, here $P\,dV$ and $V\,dP$. Relative to to the quantities of interest, Sum $dP\,dV$ can be expected to approach zero as the number of terms in the series increases as long as the total changes in P and V remain bounded. This is easy to see by assuming a uniform maximum value for dP, factoring it out of the sum, and observing the diminishing upper bound on the error caused by its omission. Therefore, the entire quantity Sum $dP\,dV$ can be referred to as an infinitesimal because it is expected to approach zero as the independent variables dV and dP are gradually diminished. In introductory calculus this idea surfaces when a definite integral is squeezed between upper and lower sums so it can be assumed that the true value of the integral lies somewhere in between the two. (Newton, 1729)

The idea of the sum of higher order terms becoming negligible as their number increases indefinitely appears in the theory of elasticity where a load-bearing member like a beam is divided into smaller elements for analysis. In this case, boundary conditions take into account the weight of a body as a whole, and only surface forces (shear, tension, and compression) are assumed to act on an elementary element. As S. Timoshenko puts it,

"Hence, for a very small element, body forces are small quantities of higher order than surface forces, and can be neglected in calculating the surface forces. Similarly, moments due to non-uniformity of distribution of normal forces are of higher order than those due to the shearing forces and vanish in the limit." (Timoshenko, p. 5)

In the theory of elasticity, a body is divided into finite elements to find an equilibrium stress distribution at a given moment in time, whereas here we are often concerned with following the behavior of a uniform sample of gas under varying conditions and disregard surface effects. G. K. Batchelor discusses the relationship between surface forces and volume forces in the context of thermodynamics. (Batchelor, p. 7)

Examples of calculations with ideal gases

The principle of path independence suggests ways to simplify the calculation of thermodynamic variable changes by breaking the path of integration into a series of short steps. Often an arbitrary short step can be represented by a pair of steps in which first one and then the other thermodynamic variable is constant.

In this way, combinations of what are called adiabatic, isothermal, isochoric, and isobaric steps can be used to calculate changes along an arbitrary path across thermodynamic variable space.

For an air sample confined by a piston in a cylinder, one elementary process is moving the piston too quickly for heat to flow so an adiabatic change occurs where $P\,dV = -\,dU$, as was demonstrated earlier in the chapter with a short computer program. This process is one starting point for the development of artificial refrigeration, since it is accompanied by a change in temperature.

At the opposite extreme, if we are patient and move the piston in slow, small steps, $P\,dV = dQ$ describes what happens while T remains close to the ambient temperature in what is called an isothermal expansion.

If we don't move the piston at all and add or remove heat, $dU = dQ$. This is called an isochoric step.

What are called isobaric steps at constant pressure can also be employed.

When combinations of step types are used, a relatively smooth path across thermodynamic variable space may be replaced by one with a sawtooth profile.

We will use combinations of adiabatic and isothermal steps to analyze the Carnot cycle, and later we will use combinations of adiabatic and isochoric steps to review Ludwig Boltzmann's explanation of entropy. Steps at constant pressure (isobaric steps) are used in the analysis of the Brayton air conditioning cycle in the next chapter.

More about air

For dry air and other ideal gases, pressure P, volume V, and Kelvin temperature T are related by the equation of state

$$PV = RT$$

that describes a sample containing *Nmol* independently moving atoms or molecules, where the symbol *Nmol* represents Avogadro's number. This number of atoms or molecules can be referred to as a mole with relatively little risk of confusion. If the kilogram is the preferred mass unit, then Avogadro's number is chosen so that the mass of a mole of oxygen atoms O is close to 16 kg; that is, close to twice its atomic number in kilograms. In the atmosphere, however, oxygen exists as a diatomic molecule, so the mass of a mole of oxygen gas is about 32 kg. Because nitrogen atoms are slightly lighter than oxygen atoms, we will use 29 kg as the mass of a mole of air, that is, the mass of a number of air molecules equal to Avogadro's number.

The mass values from a periodic table take into account the discovery of isotopes, so the values are not precise integers. Avogadro's number turns out to be about 6.02E26 molecules per kilogram mole.

For kilogram moles, the value of the universal gas constant R is about 8322 J K^{-1} kgmol^{-1} (Joules per Kelvin per kilogram mole). Writing the equation of state as

$$PV = nRT$$

allows for the situation where more or less than a mole of gas is present and also the case where a full mole of gas would not seem sufficiently homogeneous, so the sample needs to be broken up into smaller portions for analysis.

In addition to the equation of state, which describes the relationship between pressure, volume and absolute temperature for a sample of gas, the internal energy of a gas sample is also needed. The internal energy U of a kilogram mole of dry air is expected to be

$$U = 5/2 \, R \, T,$$

so the heat capacity at constant volume is

$$Cv = 5/2 \, R = 20805 \text{ J kgmol}^{-1} \text{ K}^{-1}$$

The molar heat capacity at constant volume at 298 K for N2 has been given as 20710 J kgmol^-1 K^-1, so this is a useful approximation.

The heat capacity at constant pressure is greater and given by

$$Cp = 7/2 \, R$$

since the work of expansion $P \, dV = R \, dT$ must be included.

For monatomic gases like helium and argon, $U = 3/2 \, R \, T$, and we will use this result to trace the properties of the state variable entropy to the behavior of individual gas molecules in Chapter 14. We will not discuss how the significant difference between 5/2 and 3/2 relates to molecular structure. (Born, p. 8)

Generally speaking, it is practical to perform thermodynamic experiments with gases kept at more or less constant volume. This is not the case with liquids and solids, and with these it is much more practical to measure change at constant pressure than to hold volume constant. Equilibrium between phases is studied at constant pressure so the relation between the heat absorbed by a sample and its enthalpy change

$$dQ = dH$$

will apply.

To find the specific heat at constant pressure for an ideal gas, we start from the first law of thermodynamics,

$$dQ = dU + dW$$

Applied to an ideal diatomic gas like nitrogen or oxygen, this becomes

$$dQ = 5\,R\,dT/2 + P\,dV$$

At constant pressure,

$$P\,dV = R\,dT,$$

so

$$dQ = 7\,R\,dT/2,$$

and

$$Cp = 7\,R/2 = 29085 \text{ J/kgmol/K},$$

which is significantly higher than the corresponding value for Cv.

Later there will be examples where thermodynamic data are given for a unit mass of sample rather than for a mole.

When the value for the gas constant is taken as $R = 8322$ J K^{-1}, the equation of state $PV = RT$ is very nearly satisfied by our choice of standard conditions:

$$\text{Pressure} = 100000 \text{ N m}^{-2}$$
$$\text{Molar volume} = 24.8 \text{ m}^3$$
$$\text{Temperature} = 298 \text{ K}$$

298 K corresponds to 25 C or 77 F, so it is comfortable but on the warm side. At an elevation of one kilometer above sea level, atmospheric pressure is reduced by about 10%.

Radiative transfer and the presence of water vapor must be taken into account when studying the thermodynamics of the atmosphere.

The gas laws

If one assumes that gas atoms undergo perfectly elastic collisions, then elementary mechanics shows that our expressions for the internal energy of an ideal gas and its equation of state are consistent with one another. The necessary steps of introducing vectors and taking averages complicate the proof, which is given for reference:

Elementary mechanics can be applied to atoms moving freely in three dimensions to suggest how the relationships $U = 3/2\ R\ T$ and $P V = R\ T$ arise for monatomic gases and also to show that the same constant R should appear in both. The essential idea is that the motion of an atom in an arbitrary direction can be expressed as the vector sum of simultaneous movements in three mutually perpendicular directions. Then the Pythagorean theorem is used to describe the kinetic energy KE of a single atom as the sum of three independent components arising from motion in the mutually perpendicular directions x, y, and z:

$$KE = KE_x + KE_y + KE_z = 1/2\ m\ (v_x^2 + v_y^2 + v_z^2),$$

where v represents velocity and m is the mass of an individual atom. A change in any one velocity component need not affect either of the others, so these components can represent independent atomic degrees of freedom. It is plausible to assume that each degree of freedom of each atom has a time average kinetic energy given by

$$<KE> = 1/2\ k\ T,$$

where T is the Kelvin temperature and k is Boltzmann's constant.

Then, for a mole of ideal gas atoms,

$$U = Nmol\ (<KEx> + <KEy> + <KEz>),$$

and substitution gives

$$U = 3/2\ Nmol\ k\ T = 3/2\ R\ T$$

since the values have been assigned so $R = Nmol * k$. The value of R is 8322 J kgmol^-1 K^-1, while Avogadro's number $Nmol$ = 6.025E26 kgmol^-1, and k = 1.380E-23 J K^-1. (Leighton, Appendix A)

Now the trick is to show that the same value of R that appears in the formula for the internal energy U should also appear in $P\ V = R\ T$, Charles' and Gay-Lussac's law for a mole of ideal gas.

To calculate the pressure caused by molecular collisions against the walls of a container, we use Newton's laws of motion to calculate the force on an area A of wall due to the impacts of molecules moving toward the wall and bouncing off. The pressure can be attributed to one degree of freedom of atomic motion aligned perpendicular to the surface and is independent of motion parallel to the surface.

We will sum over values of the perpendicular velocity squared to find the total pressure, so we will need to calculate the average of a function of the velocity. For this we use the convention that the average or expected value $<f>$ of a physical quantity f is the sum of its distinct values fi multiplied by the number of times $n(i)$ each value fi is observed divided by the total number of observations:

$$<f> = \text{Sum}[n(i)\ fi]/\text{Sum}[n(i)]$$

This definition is used in the following argument:

For calculating the pressure acting on an area of wall, we start with the force Fi due to the Ni atoms with perpendicular velocity in the range $v(i)$ to $v(i+1)$ that are initially included in a prismatic molar volume $Vmol$ with base area A as it begins to sweep through the position of the wall with velocity $v(i)$. The included atoms stay in step with the moving volume until they strike and bounce off the wall. The numbers of atoms entering and leaving through the sides of the volume compensate, and the two velocity components parallel to the wall do not contribute to the pressure. The sum over index i of the Ni is equal to $Nmol/2$ because we undercount atoms by a factor of 1/2 by considering only those that are initially moving toward the wall, since atoms moving away from the wall contribute equally to the internal energy.

Notice that all the volumes used for the different velocities can be thought of as superimposed only at the start of an imaginary experiment. Generally, the volumes move along one after the other in trains that have different velocities depending on the index i.

The total force acting on the wall area A is found as a sum over the velocity indices i. This will include the contributions of molecules moving with all perpendicular velocities vi, and this total force is the quantity we will divide by the area A to get the total pressure. We start by considering the train of molar volumes for a single velocity vi only and then sum over i to get the total pressure at the end.

We apply Newton's law

$$F = m\, a$$

to the case of a single atom bouncing off a surface and extend the result to a steady rain of atoms as follows: If a single collision occurs, then the push or impulse is described by the integral of force with respect to time

$$\text{Integral } F\, dt = m\, \text{delta } v = 2\, m\, vi,$$

and it is quickly over. However, for multiple impulses of short duration and a steady atomic impact rate of dni/dt due to the ni atoms per molar volume that are traveling toward the wall at vi, the persistent, fluctuating force on the area A is

$$Fi = dni/dt * 2\, m\, vi$$

The next step is to find an expression for the impact rate dn_i/dt, which is related to the quantities v_i, N_i, and $V_i = V_{mol}$. The number of collisions $dn_i(dx_i)$ produced by a movement of the ith volume a distance dx_i toward the wall is

$$dn_i(dx_i) = (A\, dx_i/V_{mol}) * 1/2\, N_i,$$

where

$$\text{Sum } N_i = N_{mol},$$

so

$$dn_i/dt = (A\, v_i/V_{mol}) * 1/2\, N_i$$

Substituting this result for dn_i/dt in the equation

$$F_i = dn_i/dt * 2\, m\, v_i$$

that was found above gives

$$F_i = (A/V_{mol}) * N_i\, m\, v_i^2$$

Then, if P_i is the pressure due to the ith velocity component,

$$P_i V_{mol} = N_i\, m\, v_i^2$$

If we sum this over the index i and use the definition of the average value given above,

$$P\, V_{mol} = N_{mol} <m\, v^2>$$

Now is the time to reap an algebraic harvest. Since all orientations of the surface on which the pressure acts are equivalent, we can write

$$P\, V_{mol} = 2/3 * 1/2\, N_{mol}\, m\, (<V_x^2> + <V_y^2> + <V_z^2>)$$

or

$$P V_{mol} = 2/3 * U_{mol}$$

From the previous result,

$$U_{mol} = 3/2\ R\ T,$$

so the molar expression

$$P V = R T$$

follows.

This important and somewhat fussy analysis is given in various sources. In place of a mole of atoms, the number of atoms per unit volume may be used. This proof is for a monatomic gas like helium or argon. The affect of molecular rotation on the result (which is needed to account for the factor 5/2 in the expression for the internal energy of hydrogen, oxygen, and nitrogen) is not treated here. (Sears, p. 318; Feynman, p. 39-2; Born, p. 3)

Clement-Deshormes experiment

Gases are so tenuous that the heat changes of a gas sample are difficult to distinguish from the heat changes of its container. We will see how this difficulty was circumvented early in the nineteenth century by the French Scientists Clement and Deshormes.

An early demonstration of cooling by adiabatic expansion is the Clement-Deshormes experiment from about 1820. This is the opposite example to the fire piston, which demonstrates adiabatic heating by compression. In this experiment, air is compressed slightly and allowed to come into equilibrium in a closed, rigid container equipped with a manometer tube partly filled with liquid so pressure changes can be measured. If the initial compression is produced by blowing into the container, around 1000 N m^-2 (corresponding to a height difference of about ten centimeters with water) is a practical starting pressure difference. Atmospheric pressure is about 100000 N m^-2, and a measured pressure difference of 1 cm of water corresponds to about 100 N/m^2.

Next, the valve is opened briefly, just long enough so the pressures can equalize. When it is closed, the menisci in the manometer loop will be at about the same level. But the volume of air remaining in the container has done work on the air it forced out, and it has cooled below ambient temperature as a consequence. It will immediately begin to warm up. Over a period of perhaps ten seconds, the pressure will visibly rise a small part of the way toward its earlier value before the valve was opened briefly and quickly closed. This step is an opportunity to observe the behavior of gas when it is warmed at nearly constant volume.

The writer remembers a version of this apparatus based on a five-gallon glass water jug of the kind once used to distribute drinking water before plastic bottles were substituted. It had a squeeze bulb with a one-way valve to produce the initial pressurization and a hose clamp to close off the rubber filling tube. When the pressure had settled to a steady level, popping and replacing the rubber stopper in the mouth of the bottle gave a crisp pressure change. Oil in a glass tube manometer mounted against a meter stick provided pressure readings.

Here the basic results were checked with a less satisfactory piece of apparatus improvised from a one-gallon plastic detergent bottle. A pour valve allowed initial pressurization and also a way to quickly open the container and close it again. A length of plastic tubing with a bore diameter of about two millimeters served for the manometer with water as the indicating liquid. The hole for the manometer tube was poked through the plastic cap provided for admitting air to the jug while pouring out detergent. For sealing the manometer tube to the cap, a hole bored undersize for a press fit and sealed with petrolatum proved best, since water-soluble household glue did not stick to the plastic cap or the plastic manometer tube. Petrolatum was used generously elsewhere. Due to an unfavorable valve geometry, the most difficult improvisation was a mouthpiece for blowing into the jug without getting a pinched tongue or a mouthful of petrolatum. A pressure shift of about 2 cm of water could be observed as the expanded gas warmed back to room temperature once air leaks had been eliminated.

The liquid-filled manometer has an interesting drawback. If the liquid is pushed entirely past the lowest point of the tube during pressurization, it will all be squirted out onto the tabletop or floor, for the volume of the tube is by design much less than that of the detergent jug.

Note that the whole amount of gas initially present doesn't remain in the container. But all parts of the sample that remain when the vent valve is closed have been expanded equally. The subsequent temperature change for the air that remains in the jug after the valve is closed and then warms to room temperature at approximately constant volume can be found as follows:

From the ideal gas law written in incremental form,

$$P\,dV + V\,dP = R\,dT,$$

it follows that

$$dP/P + dV/V = dT/T$$

as the gas cooled by adiabatic expansion warms back up to room temperature Here dP is about 2.4 cm of water, P is about 1000 cm of water, and T is about 298 K. Since the pressure recovery is approximately isochoric, $dV/V = 0$, and substitution gives $dT = 0.72$ K.

The Clement-Deshormes experiment may be the simplest way to observe the cooling effect of an adiabatic expansion, since it persists for only a few seconds in small scale demonstrations.

Loose ends

Racetrack, a physics game that is played on graph paper lined off in squares, can provide a compelling introduction to numerical integration. This game is used to introduce the concept of acceleration at the high school or freshman level. Specific information is available on the Internet.

An Internet search indicates that John G. Kemeny and Thomas E. Kurtz designed the original BASIC computer programming language at Dartmouth College in 1964.

Chapter 8
Air Cycle Air Conditioner

By using methods outlined in Chapter 7, we are now in a position to use numerical integration to predict the performance of a hypothetical air conditioner that uses air as its working fluid. Air drawn from the interior of a house can be heated by compression and then cooled close to outdoor ambient temperature by passing through a heat exchanger. Then it can be adiabatically cooled by expansion to below its original temperature and recirculated through the house to offset heating effects of sunlight and the higher temperature of outdoor air compared to indoor air.

Batches of air are carried sequentially through the steps of this cycle, and we can calculate as if the air circulates in one mole increments even though it is constantly being stirred as it passes through the heat exchanger and the house interior.

Analysis based on the first law of thermodynamics, the equation of state, and the temperature dependence of internal energy is used to calculate the cycle properties. To predict cost, we need to know the joules of electric energy needed to cool a cubic meter of air by a suitable amount below the ambient indoor temperature. This temperature difference can be measured by holding a thermometer in front of an air vent.

The power required to cool air

What shall we compare our predicted performance to? We will use data for a split cycle, two phase air conditioner that has performed well in a hot, dry climate. It is a 4-ton (nominal) unit that draws 2.4 kW of electric power and supplies 0.62 m^3 s^-1 (0.62 cubic meters per second) of conditioned air to the interior of a two-bedroom house. For this example, the power drawn was measured in the early morning of a warm July day with a stopwatch by counting rotations of the electric power meter for the residence. (Since then, household electric power consumption data have become available via the Internet.)

The air was cooled about 12 K below the indoor ambient temperature set on the thermostat: This temperature was measured at one of the outlet vents through which conditioned air entered the house interior. The total amount of air circulated per second was determined by measuring the flow velocity at the common return outlet, which measured 0.5 m^2 in area. The return grill with air filter was in the ceiling, so the velocity could be estimated by inflating a rubber balloon sufficiently so the air entering the grill would just support it. Then the balloon was dropped in still air and its rate of fall measured with a stopwatch. The value found was 1.23 m s^-1, so the estimated flow rate was .5 * 1.23 = .62 cubic meters of air per second. Then the calculation

$$(2400 \text{ W})/(.62 \text{ m}^3 \text{ s}^{-1}) = 3871 \text{ W s m}^{-3}$$

gives 3871 joules as the energy needed to cool a cubic meter of air 12 K below room temperature, so 3871 watts of electric power per cubic meter of air cooled per second can be used as a reference value. 12 K will be used as a typical value for the temperature difference between indoor air and air coming from a cooling duct.

A calculation based on values found on the Internet suggests 3700 watt per cubic meter per second as the typical power requirement for producing cooled air with a home air conditioner.

There are about 24.8 cubic meters in a kilogram mole of air, and we will use the rough estimate of

$$3871 * 24.8 = 96000 \text{ J/kgmol}$$

as the energy requirement for cooling air by 12 K with this commercial unit.

How much heat would be needed to supply this cooling if chilled water flowing in a counter-flow heat exchanger were used to cool the circulating air on its way to the vent instead? We use the specific heat Cp of air at constant pressure:

$$Cp = 7*R/2 = 7*8322/2 = 29127 \text{ J kgmol}^{-1} \text{ K}^{-1}$$

Then

$$Cp * deltaT = 29127 * 12 = 349524 \text{ J/kgmol}$$

is the energy requirement for cooling a kg mole of air by 12 K with this unit. Thus about 349524/96000 = 3.64 times more energy would be be budgeted to cooling the house if previously chilled water were used to absorb the heat directly.

This result may astonish because the heat absorbed by the liquefied refrigerant as it evaporates appears so much greater (by a factor of 3.64) than the work to condense it. What has become of the first law of thermodynamics, of the equivalence between heat and work? Here there is also a compensating flow of heat from the refrigerant to the outdoor air to take into account.

The beneficial factor of 3.64 obtained by dividing the power for direct cooling by the electric line power is known as the COP or coefficient of performance for the air conditioner.

The test reported here was made early in the day, and the COP of the system would have been less if it had been measured in the afternoon. (Ananthanarayanan, p. 50)

Now we are ready for the principal task of this chapter, which is to calculate the energy needed to cool the house assuming an air expansion engine working on the Brayton cycle is used to cool the air.

The Brayton cycle

The Brayton cycle is a convenient way to use adiabatic expansion to produce refrigeration. Air is compressed and then cooled in an outdoor heat exchanger so it will be cooler than the indoor ambient temperature after a subsequent expansion. This same thermodynamic cycle was operated in reverse to provide power in an early form of internal combustion engine. The Brayton heat engine cycle thrives today in jet engines used to propel airplanes. (Sandfort, p. 257)

In our application of the Brayton cycle, room temperature air from inside the house is compressed adiabatically using electric power, cooled at constant pressure in an outdoor heat exchanger, expanded adiabatically to further cool it below room ambient temperature and produce recoverable work, and finally warmed at constant pressure as it circulates through the house to deliver air conditioning and complete the cycle. An internal combustion engine typically operates on an open cycle as it repeatedly draws in fresh oxygen and exhausts combustion products to the atmosphere, but for air conditioning, the air is assumed to be recirculated continuously through the house, which need not be tightly sealed.

(The Brayton cycle can serve either for an engine or a refrigerator, but thermal gradients must be turned around to make the switch, so it is not reversible in the same way a Carnot engine used for making thermodynamic measurements would be. This will be discussed in Chapter 10.)

Predicted Brayton cycle performance

Numerical integration can be used to predict the performance of a hypothetical air conditioner that uses dry air as the working fluid and operates in a Brayton cycle. Ambient house air is first heated by adiabatic compression to well above the outdoor temperature and then cooled at constant pressure while passing through a heat exchanger located outdoors. This heat exchanger needs to cool the air coming from the compressor down close to the outdoor ambient temperature. Then the compressed air is further cooled by adiabatic expansion in the cylinder of a compressed air engine and exhausted into the house to cool the indoor air by a second heat exchange at constant pressure. Work from the expansion engine compensates for some of the power needed to run the compressor.

Over time, equal amounts of air pass through the compressor and the expansion engine. It is convenient to think of the compressor piston and the expansion engine piston as working off the same crank shaft. The compressor valves are pressure actuated, and the timing of the cam-operated engine valves, which are similar in function to those of a steam engine, must be controlled to suit the requirements of the cycle as a whole.

The fixed reference state of the air can be assigned as it enters the return duct to the compressor inlet at indoor ambient temperature and pressure.

Calculating the pressure versus volume diagram for a one mole sample of air passing through the entire loop is the starting point, even though air samples would mix as they pass through the outdoor heat exchanger and especially the house interior. It is convenient to assume that one mole of air is forced into the outdoor heat exchanger by each compressor stroke.

To calculate the energy cost of air conditioning, we need to make assumptions about the cycle conditions. We will assume the outdoor air temperature is 314 K (106 F or 41 C) and that air from inside the house is adiabatically heated by compression from 298 K to a temperature of 344 K before entering the heat exchanger, where it cools to 329 K. Having an ample temperature difference of 15 C between the outdoor air temperature and the temperature at which the refrigerant leaves the high temperature heat exchanger assures that a practical heat exchanger can be used and that excessive power will not be needed to run the cooling fan. (Notice that cooling fan power is included for our reference commercial unit but not for the air cycle design we are analyzing in this chapter. A total fan power of around a kilowatt is expected for producing cooled air at a rate of about one cubic meter per second.)

Cooling by adiabatic expansion of the compressed air is from 329 K to 285 K, but 298 K remains the preferred reference point in the analysis: When the air conditioner is operating under steady load, it will be assumed that house air enters the compressor intake at a pressure of one atmosphere and a temperature of 298 K.

Air is drawn into the compressor cylinder through a pressure-activated intake valve, and minimal head space is allowed at the top of the stroke. Therefore the initial volume of a sample of compressed air entering the much larger volume of the outdoor heat exchanger equals the cylinder volume above the piston face at the moment the pressure equalizes and the exhaust valve opens. The final volume for an equivalent sample of air as it leaves the heat exchanger at the same pressure but at a lower temperature is determined when the intake valve of the expansion engine closes and adiabatic expansion begins. The system must be designed and regulated to match the initial and final volumes assumed for a typical sample of air, which in this case is a kilogram mole.

Next we present the BASIC program used to analyze what happens as a mole of air circulates through a Brayton air conditioner. First the initial values of pressure, volume, and temperature are given, and a suitably small volume increment absdV is selected.

Here is the start of the program, which is named Brayton 230103:

```
print "Brayton 230103"
P=100000
V=24.8
absdV= .001
T=298
R=8310
U=0
Qh=0
Qc=0
```

W=0
S=0

Now we imagine that the lightly spring-loaded compressor intake valve snaps shut at the end of the intake stroke and the compression stroke begins. This stroke will end when pressures equalize between cylinder and heat exchanger and the exhaust valve opens automatically due to the pressure differential changing sign. Here is the next section of code. It is the same as for the adiabatic expansion in Chapter 7, except we are compressing the air rather than expanding it:

(Start adiabatic compression and warming.)

```
while V>17.3
dV= -1*absdV
V=V+dV
dU=-1*P*dV
dT=2*dU/R/5
U=U+dU
T=T+dT
P=R*T/V
dW=P*dV
W=W+dW
wend
```

The final volume of 17.3 m^3/mole has been chosen to give the temperature change desired during adiabatic compression.

After adiabatic compression
P = 165322
V = 17.3
T = 344
W = -959302

The air has been compressed to 1.65 atmospheres pressure and heated to 344 K. This is hot enough so that there will be plenty of temperature difference with the outdoor air to cool the circulating air to 330 K while remaining at 1.65 atmospheres pressure. This is done in a heat exchanger located outside the house. Almost 10^6 joules are required to compress a mole of air to the pressure required.

The next step is to calculate what happens as a mole of air passing through the heat exchanger is cooled to 330 K. The volume change is from the volume it occupies in the compressor cylinder as the exhaust valve opens to the volume it occupies in the expansion engine cylinder as the intake valve closes. (To calculate the cycle work, it doesn't matter that the mole of air at this stage of the cycle is different from that at the first.) Now heat is flowing out of the circulating gas and into the atmosphere through the walls of the outdoor heat exchanger tubes, and the amount of heat lost per mole Qh must be calculated using the first law of thermodynamics:

```
'Start isobaric cooling.
while T>330
dV=-1*absdV
V=V+dV
Tr=T
T=P*V/R
dT=T-Tr
dU=5*R*dT/2
U=U+dU
dW=P*dV
W=W+dW
Qh=Qh+dU+dW
S=S+(dU+dW)/T
wend
```

Because heat is flowing, the molar entropy change S for the circulating air is also calculated.

After isobaric cooling
P = 165322
V = 16.587
T = 329
Qh = -412561
W = -1077177
S = -1224

The temperature of the compressed air falls steadily while passing through the heat exchanger, here from 344 K to 329 K. 329 K is significantly hotter than 314 K, the outdoor air temperature anticipated on a hot summer day, so we can safely assume there will be an adequate temperature difference to drive the heat exchange when outdoor ambient air is used to cool the compressed air as it passes through a set of tubes cooled by circulating outdoor air. This heat exchanger performs the same function as the outdoor heat exchanger of a typical split air conditioning system for residential use. (The term split indicates the contrast with window air conditioning systems, which package the entire system in one box, instead of divided between the yard and the attic or a utility space.)

Notice there is net work done on each gas increment as it passes through the heat exchanger even though the volume of the heat exchanger is constant. This is because a mole of hot gas pushed out of the compressor occupies a larger volume than a mole of gas that enters the expansion engine during the intake portion of its stroke. This is what adds an additional component to the total cycle work.

Now we can introduce the expansion engine. It is similar to the compressor with minimal head clearance at top dead center. If the intake volume can adjust for the loss of volume as a mole of air passes through the heat exchanger, then the expansion engine might operate off the same crankshaft and flywheel as the compressor and automatically recover mechanical energy. The big difference between the engine and compressor is that cams must operate the valve gear to produce the following effect:

The intake valve for air coming from the heat exchanger opens at near top dead center with minimal clearance volume and remains open for about seventy percent of the power stroke. The intake valve closes, and adiabatic expansion takes place to near bottom dead center. Then the exhaust valve opens to the house interior and remains open until just before top dead center. (To minimize exhaust noise, the exhaust valve should open as the pressure in the engine cylinder falls close to house ambient pressure.) The effect of the adiabatic expansion is calculated as follows:

```
'Start adiabatic expansion.
while P>100000
dV=absdV
V=V+dV
dU=-1*P*dV
U=U+dU
dT=2*dU/R/5
T=T+dT
P=R*T/V
dW=P*dV
W=W+dW
wend
```

(This the same process as in the Chapter 7 adiabatic expansion example.) Even though the exhaust valve should open when there is near zero pressure difference between the house interior and the cylinder interior, a cam mechanism is needed to hold this valve open during the exhaust stroke and then close it just before the intake valve opens. The program output shows that the exhaust air is indeed cooler than the air that originally entered the compressor intake at the same pressure:

After adiabatic expansion
P = 99999
V = 23.753
T = 285
W = -159967
S = -1224

The cooled air next spreads through the house interior. It blends with the house air and keeps its temperature comfortable except for a drafty region in the vicinity of the exhaust vent. This is described by the last part of the cycle:

```
'Start isobaric expansion.
while T<298
dV=absdV
V=V+dV
```

```
Tr=T
T=P*V/R
dT=T-Tr
dU=5*R*dT/2
U=U+dU
dW=P*dV
W=W+dW
Qc=Qc+dU+dW
S=S+(dU+dW)/T
wend
```

The final output shows the outcome for a mole of air passing through the cycle: A typical sample has very nearly returned to its initial state, and the sum of the heat lost to the outdoor air Qh < 0 and the heat absorbed from the house interior Qc > 0 is nearly equal to the mechanical work W required to accomplish a one mole cycle:

After isobaric expansion.
P = 99999
V = 24.764
T = 298
U = 24.5524572
Qc = 353848
W = -58868
S = -11.8434078

Qh+Qc = -58713
W = -58868
S = -12

The final three results (which are not computed directly by the original program) show that the first law of thermodynamics is obeyed by a typical mole of air passing around the cycle. In addition, the total entropy change S for a mole of air is near zero, consistent with the hypothesis that entropy as well as internal energy is a state property.

Notice especially how the state properties of pressure, volume, temperature, and entropy return to their initial values at the end of the cycle while large quantities of heat and work are absorbed or produced in overall agreement with the first law of thermodynamics. William Thomson (aka Lord Kelvin) is credited with noting this very significant result. (Sandfort, p. 87)

Promise and problems

Our Brayton air conditioner theoretically requires $W = -58868$ J/kg mole to cool air 12 C below room ambient temperature. Since we calculated earlier that $Cp * deltaT = 349524$ J/kg mole for cooling air 12 C, we conclude

$$COP(Brayton) = 349524/58868 = 5.9$$

which is greater than the value of 3.6 measured for the commercial unit. This result is encouraging for two reasons: First of all, we have neglected to include the heat exchanger fan power for the Brayton system. A second reason for the favorable theoretical performance of the Brayton refrigerator is that we have simplified the system to the utmost by assuming the expansion engine exhausts directly to the house interior. This is problematic because of the noise that would result and because a mist of lubricating oil might enter the house along with cool air. It is likely a second heat exchanger would be arranged in the plenum of a cooling duct system as is done with conventional two-phase air conditioners. In this case, some additional expansion would be required to cool the engine exhaust air closer to the freezing point of water before it entered the indoor heat exchanger, and this would boost the power required.

One result of the Brayton cycle analysis is discouraging. Suppose the compressor displacement is one liter, so the piston sweeps through roughly the volume of a canning jar at every stroke. Since a liter is 1/1000 of a cubic meter, something like 600 piston strokes a second would be needed to cool a home like the one for which air conditioner performance was measured, and this does not appear to be practical. Bulky machinery is needed to make this approach work. Sealing the expansion engine-compressor system by adding a second heat exchanger and using gas at a much higher pressure might help solve the problem. This could also make the system less noisy and easier to live with, and problems associated with moisture condensation might be avoided.

A control system would be required for the Brayton air conditioner. Perhaps reducing power to the cooling fan in the outdoor heat exchanger on cool days could be used to maintain operation on the precise cycle selected. But there is no need to address this design problem here.

There is one significant puzzle that may pass unnoticed. The Brayton cycle machinery could be reconfigured to operate as an engine and produce power. This is not the case with the standard system operating with the refrigerant passing back and forth between a liquid phase and a gas phase. It can be used as an air conditioner in summer or a heat pump in winter, but it cannot serve as a source of mechanical power.

Chapter 9
Children of Prometheus

The development of the steam engine prepared the way for subsequent work on refrigeration machinery. Both depended on the scientific knowledge and technical skills that evolved together as Europe emerged from the Middle Ages. Development of transparent glass tubing made possible improved understanding of the behavior of liquids and gases. Torricelli invented the compact mercury-in-glass barometer, and Boyle used mercury in glass tubes to discover the principle that the product of pressure and volume are constant for air kept at constant temperature. The synthesis of ether was discovered. Technologies previously reserved for the production of weapons and luxury goods assumed broader roles. This progress contrasted with episodes of violence like the persecution of witches and European wars of religion.

In the seventeenth century, scientists learned that steam condensing in an airtight container could suck up water from a well through a pipe. This seemed like the basis for a practical pumping process even though water could only be lifted about ten meters in a single stage. By 1712, the Englishman Thomas Newcomen had eliminated this problem by replacing the single vessel that was alternately filled with steam and water with separate steam and pump cylinders that were fitted with pistons. His engine pumped water continuously, stroke after stroke, when supplied with steam and cooling water.

British technical historian John Farey described Newcomen's engine in remarkable detail a century after its introduction. (Farey, 1827)

British coal mining created the demand that made Newcomen's invention practical. (Carnot, p. 2) If a deep coal mine needed to be cleared of water, the pump cylinder could be set at the bottom of the shaft at the end of a long pump rod. Coal from the mine could be used as fuel. Water from the mine was available to fill the boiler and cool the cylinder to create the necessary vacuum; however, surface water from a purer source with less dissolved mineral content was preferred if it could be obtained.

Newcomen engines pumped water to turn mill wheels and drive textile machinery but were not applied to ship propulsion with lasting success.

Wrought iron, cast iron, and wood were important construction materials for machines at the time, and low-pressure boilers could be made from copper and even lead. If we consider the skill required to make a flintlock musket or a clock, then the Newcomen engine does not appear too much of a challenge. Size was a problem though, for the cylinder could be a meter in diameter by three in height. It was set upright, the open end up, and needed to be held down by sturdy fastenings or it would lift off its foundation as a vacuum formed under the piston.

A pivoted beam connected the piston rod to the pump rod, which was offset from the axis of the piston stroke. This beam resembled the ones still used in oil well pumps except that it was built from timbers held together with iron fasteners.

The piston worked up and down. Farey reports that hemp packing with tallow was placed on a ledge running around the outside of the piston, and that this packing was compressed against the cylinder wall by iron weights cast in the form of segments of a circle. Water and tallow could be added from above to maintain the seal and prevent air from leaking into the cylinder when a vacuum formed.

How well might messy masses of plant fiber, hot grease, and water control and confine steam? Experiments with cotton wads in plastic drinking straws can supply a little information. Adding petrolatum or vegetable oil significantly improves the effectiveness of the seal when a chop stick or skewer is used to push down on a wad and pump bubbles out of the end of a straw at a depth of a few centimeters under water. (It has been suggested that a leather cup with water on top could have been used to seal the gap between piston and cylinder in Newcomen's engine. Farey does not appear to confirm this, and pliable leather, which works so well in a bicycle tire pump, can become hard and stiff when exposed to steam.)

At the start of the up stroke with the piston close to the bottom of the cylinder, steam at slightly above ambient pressure was admitted to the cylinder to purge air and water through a one-way valve. Then the weight of the pump rod helped pull the piston up through the cylinder while steam continued to flow in. At the top of the stroke, the steam intake was closed, and condensing water was vigorously sprayed into the cylinder through a nozzle. For maximum work output, as much steam as possible needed to be condensed before the piston was far into its descent. Water for cooling was pumped and stored at a location high enough above the injection nozzle so it would rush briskly into the cylinder and promptly produce suction as the injection valve opened. (An Internet search may turn up an indicator diagram for this cycle.)

At the bottom of the power stroke, pressure that had been building in the boiler served to eject air and accumulated water through the one-way valve while the piston was reversing its motion. For this, the boiler pressure needed to increase to slightly above atmospheric pressure. Then the rate of steam production could limit the speed at which the piston rose higher in the cylinder.

One can test the principle of the Newcomen engine with a stemmed wine glass and a teakettle. Water is brought to a boil with the whistle removed so steam can escape without producing turbulence, and a dish of cold water is set next to the kettle. The glass is inverted over the spout so escaping steam can rise into it and displace the air. Water droplets will condense to show this is happening. Hands must be kept clear of the steam to avoid a painful demonstration of latent heat of condensation. When the glass is hot, it is kept inverted and quickly dipped rim down into the water in the dish. The steam will condense, and water will gradually rise perhaps halfway into the glass when all the steam has cooled and turned back into water. Air that became mixed with the rising steam is what remains above the water level.

This demonstration is an example of how mass-produced, commonplace goods can be transformed into tools for the advancement of understanding beyond what prehistoric people could achieve. (Derry, Ch. 25)

Britain's special needs and abundant coal resources favored the development of the Newcomen engine, even though its fuel economy was poor: To do the maximum amount of work, enough water had to be sprayed in at the start of the power stroke to cool the entire interior of the cylinder from the temperature of boiling water to ambient temperature. Then, during the subsequent intake stroke, the interior surface of the cylinder and any water remaining were heated back to boiling temperature again. A way to avoid alternately cooling and reheating the cylinder was needed to save fuel. Later in the eighteenth century, James Watt famously added a separate condenser so the interior of the engine cylinder did not need to be sprayed with cold water during every power stroke. This significant improvement made the steamboat practical, and by around 1800 the world was set on a path of startling and highly visible technical transformation.

Watt still condensed the steam with a jet of water instead of a heat exchanger analogous to the condenser coils used with a modern air conditioner. He had saved himself a problem, because a surface condenser might have approached the boiler in size. American refrigeration pioneer Dr. John Gorrie, who invented an air cycle refrigerator in the 1840s, apparently followed Watt's example and relied on a jet of water to cool compressed air before expanding it. (Sandfort, p. 167)

Farey reports that Watt sealed his piston using the same general approach as the one applied with Newcomen's engine. However, the hemp gasket saturated with tallow was held in place by bolts and a clamp ring rather than weights. The gasket was made stiffer by excluding water from the engine cylinder, so maintaining a seal became more difficult in cylinders manufactured to the original low standard of precision that was adequate for Newcomen engines. Water on top of the piston no longer helped seal the gap between piston and cylinder, and unevenness in the cylinder wall made it more likely that contact with the greased gasket would be lost so steam could blow through the gap.

British manufacturer John Wilkinson perfected a boring machine combining the functions of a lathe and a milling machine that was adapted to the task of accurately shaping the interior surfaces of rough steam engine cylinder castings. This machine was subsequently employed to supply cylinders to the firm of Bolton and Watt. (Derry, p. 350) Machines that perform a similar function are still used to refinish worn automobile engine cylinders before new pistons are installed.

Watt's and Newcomen's configuration with steam at close to one atmosphere pressure and condensation at ambient temperature was the original one. By around 1810, there was another approach in use. Water was boiled at a pressure of several atmospheres, and steam was exhausted directly to the atmosphere the way steam railroad locomotives still do today.

The simplification of doing away with the condenser in high pressure engines came at a cost: With higher pressure came greater risk of a lethal boiler explosion. Carnot's work would eventually make it easier to weigh possible fuel savings against the difficulties associated with each configuration. Carnot recognized that the temperature at which water condensed and the temperature at which it boiled to steam were critical variables in determining how much work could be done. The big temperature difference between the boiling temperature and the temperature of the fire under the boiler served to "push" more heat through the boiler wall and evaporate water more rapidly. This was an important consideration for railroad locomotives. (Ripper, Ch. XXV)

Metallic packing in the form of piston rings was also introduced at about the same time as high-pressure steam. The first ring systems had multiple segments; the one-piece spring rings we rely on today in automobiles were introduced later by Ramsbottom. Metallic rings were promptly adopted for railroad locomotives. J. E. Gordon states that George Stephenson used metallic packing rings in his revolutionary *Rocket* locomotive. (Gordon, p. 246) John H. White, Jr., indicates that by the 1830s only metallic piston packing was used in American railroad locomotives. (White, Jr., John H., p. 207)

However, the obsolete system of soft packing used in Newcomen engines and by Watt should not be dismissed as totally without merit. Louis C. Hunter reports that greased hemp cylinder packing remained in use on Ohio and Mississippi River steamboats for over forty years until 1862 (Hunter, p. 166) Powerful engines and hulls approaching 150 feet in length were needed to steam upstream against the Mississippi current, so the method must have had its merits. (Hunter, p. 87) Perhaps riverbank stops to adjust cylinder packing seemed preferable to increased mechanical complexity and first cost. Hunter suggests that fires, boiler explosions, and cholera could easily have been more concerning. When Fanny Trollope traveled up the Mississippi by steamboat in 1828, she had plenty to write about besides the celebration of ordinary workaday achievements. (Trollope, c. 1830)

From the time of James Watt on, engineers and mechanics could see how their steam engines were working in intimate detail with the help of the indicator, a pressure-actuated mechanism that generated pressure versus volume plots in real time when mounted on an engine frame and coupled by an actuating mechanism to the piston cross head and by a steam tube to the cylinder headspace. (Ripper,

Ch. XV) For those who worked with machines, there was gain and loss as internal combustion swept steam away--gain because of the control over the cycle provided by adjusting the spark timing and loss because of reduced knowledge about what was happening inside the cylinders of a misbehaving engine.

Understanding how steam engines were designed and built provided a foundation of knowledge for the developers of refrigeration machinery when the demand for ice and better ways to preserve food increased in the nineteenth century. (Derry, p. 698)

Today the nineteenth century may seem like a distant past despite the steady stream of remarkable innovations it brought forth. The literature of the time suggests that major shifts in attitudes and preoccupations have taken place. Consider, for example, how technical expert E. P. Watson promoted his book on engineering practice:

"Conceive, then, the delay and hindrance caused by neglect or mismanagement. Let one man in a district of ten square miles be five minutes late in starting his machinery in the morning; and reflect if there be one in such a predicament within the limit prescribed, what a loss will ensue to the country at large, through all its towns and cities." (Watson, p. 145.)

Contrary to what one might expect, Watson's book offers a wealth of tantalizing information of the kind that is prized by hobbyists and other enthusiasts.

Chapter 10
Carnot's Engine and the Second Law of Thermodynamics

In Chapter 8 we used theoretical knowledge of the heating and cooling produced by adiabatic volume changes in conjunction with the first law of thermodynamics to design an air conditioning system with air as the working fluid. The predicted coefficient of performance (COP) of the air cycle system compared plausibly with the measured COP for an actual air conditioner. However, we are still far from our goal of understanding the air conditioners in general use because they rely on the evaporation of a liquid refrigerant at constant pressure to produce cooling and not on the adiabatic expansion of an ideal gas.

The work with the air cycle system is not wasted, however, because the Brayton cycle of Chapter 8 can be modified to produce the standard engine cycle to which the performance of other engines and refrigeration systems is compared. This approach was proposed by the French Physicist Sadi Carnot early in the nineteenth century. As discussed in the previous chapter, steam engines were on the rise. Engineers worried about losses due to steam condensing in engine cylinders and sometimes wondered if another working fluid other than water and steam might be superior. Generally speaking, they were already headed in the right direction, and Carnot's work received little attention at first. (wikipedia en francais, Sadi Carnot)

To understand what limited engine performance, Carnot focused on a significant distinction. He realized the red heat of the fire under a boiler served mostly to push heat through the boiler wall to produce steam. What limited the amount of work the steam produced was the difference between the temperature at which water boiled and the temperature at which it condensed. The greater the temperature difference here, the more work could be done from a given amount of fuel.

To better quantify steam engine performance, Carnot came up with a design for a perfectly reversible heat engine to which the performance of steam engines (and eventually internal combustion engines) could be compared. A hot air engine suited his purposes best, but it differed from the Brayton cycle machine in Chapter 8 because a single sample of gas would be permanently retained in a cylinder fitted with a piston. The goal was to produce a thermodynamic cycle where all the increments of heat and work would reverse precisely when the engine was run backward as a refrigerator. To achieve this, the gas was to be first heated at constant temperature by placing the cylinder in contact with a source of heat at high temperature and then cooled at constant temperature by placing the cylinder in contact with a colder heat sink. For this engine to have the same performance characteristics when reversed and operated backward as a refrigerator, the heat reservoir temperatures would need to nearly match the cylinder gas temperatures during heat exchanges. Thus the engine would have to run so slowly that its power output (but not its work output) would be negligible for practical purposes. In between the heat exchanges at the two temperatures selected, the temperature of the cylinder gas would be adjusted first by adiabatic expansion and then by adiabatic compression to complete the cycle.

By proceeding in this way, Carnot created an engine design that came with a built-in test: If the engine was set up properly and run slowly enough, its performance as a refrigerator would be equal and opposite to its performance as an engine.

The reason a quick-running cycle is unsuitable for Carnot's purpose is that thermal gradients must reverse during the switch-over from operation as an engine to operation as a refrigerator. An Internet search suggests Sadi Carnot was the first to take this effect into account when he introduced the concept of a reversible heat engine, while Dr. Gorrie's air cycle ice maker may be the earliest practical example of cooling by means of a hot air engine variant running in reverse.

By a stroke of insight, Carnot realized that any chemically inert gas could be used as a working fluid in his engine. With a reversible engine, the cycle work could not depend on the choice of gas. Otherwise, two engines using different gases could be synchronized to convert heat from a single reservoir entirely into work without the need for a colder reservoir. One may imagine how many cooling towers such a scheme could clear from the landscape! For an engine to work in cycles without discharging heat to the environment would be comparable to the reverse of Benjamin Thomson's friction experiment. It is in the same category as the fantastic idea that pushing a drill bit against something hot might cause it to rotate spontaneously.

Practical engines working between the same temperature extremes as a Carnot test engine could be expected to be less efficient, so his engine would prove a reliable standard for establishing upper limits to performance. Thus Carnot had also found the way to produce an ideal thermometer that would require only one calibration reference.

Unfortunately, Carnot pictured equal quantities of heat being absorbed and discharged at two distinct temperatures during the operation of his engine, and he had difficulty analyzing the isothermal expansion and contraction of the gas sample. (The isothermal expansion is produced by moving the piston as required to prevent a significant temperature rise while the cylinder is being heated, and conversely for cooling.) In our picture of the process today, the difference between the amounts of heat absorbed and emitted is equal to the cycle work since the internal energy of the cylinder gas is restored to its initial value at the end of a cycle.

An indication of Carnot's difficulty is found in the description of his engine when he indicates that he does not know (and presumably has not been able to measure) the precise rate at which heat is absorbed from a heat reservoir during an isothermal expansion. Today we easily estimate the amount of heat needed for isothermal compression of an ideal gas. But the relationship was unclear to Carnot: "*Nous ignorons quelles lois il suit relativement aux variations de volume*," he wrote. (Carnot, 1824, p. 17)

Because of a mistaken assumption, Carnot's work remained incomplete. In the state he left it, his engine design could be applied to measure performance limits for heat engines working between different temperature extremes, but the limits could not yet be calculated from theory. Today it seems natural to assume that the work W done by one complete cycle of an ideal heat engine is given by

$$W = Qh + Qc,$$

where Qh is the amount of heat absorbed from the high temperature thermal reservoir and Qc is the amount of heat discharged to the environment. The quantities Qh and Qc in turn obey

$$Qh/Th = - Qc/Tc$$

(Note that heat leaving the high temperature reservoir can be taken as negative and heat entering the low temperature reservoir can be taken as positive. Or the heat entering the cylinder gas can be positive and that leaving can be negative.)

In his thermodynamics text written decades after Carnot's death, Max Planck emphasizes that Carnot's discovery can be expressed as:

"*It is impossible to construct an engine that will work in a complete cycle and produce no effect except to raise a weight and cool a heat reservoir.*" (Planck, *Thermodynamics*, p. 102)

In a footnote, Planck adds that this statement of the second law of thermodynamics coincides essentially with the starting point chosen by the scientists Rudolph Clausius, William Thomson, and James Clerk Maxwell who came after Carnot. It might seem possible to further cut the word count and say:

There is no way to take heat from a single source and convert it completely to work.

However, the idea of a complete thermodynamic cycle is an essential feature of Carnot's discovery.

Acceptance of the first law

Let us consider how people progressed from thinking

$$\text{Sum } dQ = 0$$

in the 1820s to using

$$W = Q_h + Q_c$$

to give the work produced by a cycle of a Carnot engine by about 1850. During this interval, the concept of internal energy U became better understood, so the formula

$$dQ = P\,dV + dU,$$

which we introduced in Chapter 5, was at last used to fully analyze the details Carnot's cycle. We will use material presented by John P. Sandfort to call special attention to work of major contributors Thomson, Clausius, Joule, and von Mayer:

In his book *Heat Engines*, John F. Sandfort quotes both the British scientist William Thomson (later Lord Kelvin) and the German scientist Rudolph Clausius to show that by around 1850 they were coming to terms with the first law of thermodynamics in the form we know it today and applying it to heat engines. First, we hear from William Thomson:

"When equal quantities of mechanical effect are produced by any means whatever from purely thermal sources, or lost in purely thermal effects, equal quantities of heat are put out of existence or generated." (Sandfort, p. 87)

And next from Rudolph Clausius:

"On a nearer view of the case, we find that the new theory is opposed not to the real fundamental principle of Carnot, but to the addition 'no heat is lost,' for it is quite possible that in the production of work, both may take place at the same time; a certain portion of heat may be consumed and a further portion transmitted from a warm body to a cold one, and both portions may stand in a certain definite relation to the quantity of work produced." (Sandfort, p. 85)

But what about internal energy? In these two passages, Thomson and Clausius can omit mention of internal energy if they are talking about complete thermodynamic cycles of samples of gas that return to an initial state at the end. The problem of internal energy was addressed by two other scientists who were also picking up where Carnot left off:

James Prescott Joule of Manchester completed Benjamin Thompson's observations by measuring with precision the thermal equivalent of work when it is converted to internal energy by friction. (Sandfort, p. 77)

German physician Julius Robert von Mayer was an early exponent of the principle of the conservation of energy passing through diverse forms. He measured the mechanical equivalent of heat and enhanced understanding of the nature of internal energy by studying the ratio of the specific heat of air measured at constant pressure to the value measured at constant volume. (Sandfort, p. 76)

By 1875 textbooks could be published that confidently referred to heat as a mode of motion. (Youmans, p. 63)

A computer model of the Carnot cycle

For his theoretical research, Carnot specified an engine that absorbed and emitted quantities of heat Qh and Qc at fixed temperatures Th and Tc respectively. Even though Watt's powerplants boiled water at one temperature and condensed it at another, Watt steam engines could not quite satisfy Carnot's needs and serve as a mathematical model, partly because liquid water was heated at constant pressure from ambient to the boiling point after being injected into the boiler.

Carnot turned to an ideal gas like air for the working substance in his theoretical engine. Heat could be added and removed almost reversibly during slow isothermal expansions and contractions of the working gas sample in the cylinder, and the necessary temperature changes could be accomplished by adiabatic expansions and contractions.

However, after developing a theoretical model for a reversible engine with optimized efficiency, Carnot could not define absolute or Kelvin temperature starting from the relationship

$$Qh/Th + Qc/Tc = 0,$$

where $Qh < 0$ is the heat leaving the hot reservoir, $Qc > 0$ is the heat entering the cold reservoir, and Th and Tc are the respective temperatures. This is because he did not apply the first law of thermodynamics in the form

$$W = Qh + Qc$$

to determine the work performed by a complete cycle of the gas in the engine cylinder. (We try to keep with the convention that heat entering a reservoir is positive and heat leaving is negative. This can fight against the instinct that we should get positive work from an engine.)

Here we will go beyond what Carnot was able to do and present a computer program that uses the first law of thermodynamics to model a Carnot engine cycle using the equation of state and the internal energy expression for an ideal gas. The calculation confirms that the work for a whole cycle obeys

$$W = Qh + Qc$$

with

$$Qh/Th + Qc/Tc = 0$$

What might an actual realization of Carnot's ideal engine be like? Carnot simplified his model and his calculations by substituting air for steam and eliminating the phase change. He considered the amounts of heat drawn from one thermal reservoir and rejected to another by a sealed engine cylinder filled with a fixed amount of ideal gas. The heat reservoirs might be imagined as two large, well-insulated tanks of water, one at 100 C and the other at 25 C. Neither fire nor a separate supply of cooling water would be needed in the few cycles it would take to measure the efficiency of his engine: openings in the insulation could be made for heat transfers and closed again. Adiabatic expansion and compression could be relied on to change the temperature of the cylinder gas without any heat needing to flow at an intermediate temperature.

(The conditions required for a Carnot cycle are difficult to produce in the laboratory because the gas in the engine cylinder must. be insulated from its surroundings during the adiabatic expansion and compression. The Clement-Desormes experiment described in Chapter 7 is much easier to carry out and demonstrates the same thermodynamic principles in action.)

Now for the calculation of a Carnot cycle using reservoir temperatures that approximately apply to Watt's steam engine. It is important to emphasize: Over a whole number of cycles, the state of the cylinder working gas sample always comes back to what it was at the start. Without the heat transfers from and to the all-important heat reservoirs, it is as if nothing had happened! Despite the ingenuity that has gone into designing engines with their valves and seals and shining shafts, in the end what counts is the heat that disappears on the way from the hot reservoir to the colder reservoir: Its equivalent appears as work done by the engine.

We will calculate for an engine gas sample consisting of one mole of air using $R = 8310$ J kgmol^{-1} K^{-1}. The compression stroke will start with the air at standard temperature and pressure, $T = 298$ K and $P = 100,000$ N m^{-2}, so the initial volume is $V = 24.8$ m^3. We will carry out an adiabatic compression to heat the air from 298 K = T_c to 373 K = T_h in order to approximately model Watt's cycle without the complications of preheating feed water and allowing for phase change during subsequent expansion.

In the following Liberty BASIC program 211129 Carnot, break points for looping are controlled using the while(condition)/wend command pair. The computation is divided into sections that include the four different cycle sections (adiabatic compression, isothermal expansion, adiabatic expansion, and isothermal compression), while output values and comments are given at break points in the program listing.

The program also gives the change in the engine gas entropy S, which comes out close to zero at the end of the cycle. The engine gas internal energy U is assumed to be given by $U = 5/2 \, R \, T$.

In the program and calculation printouts, variable symbols are printed in a line followed by lines giving output values row by row.

This is done to satisfy text formatting requirements for ebook publication. Code sections and the corresponding printouts are grouped together: Commands and variable names appear in the code sections and numerical values in the output sections that follow.

The first section of the program assigns and prints variable values:

```
print "211129 Carnot"
print "Liberty BASIC"
P=100000
V=24.8
absdV= .01
T=298
R=8310

Qh=0
Qc=0
W=0
S=0
print

print "Initial values"
print "P V T Qh Qc W S"
print int(P)
print V
print T
print int(Qh)
print int(Qc)
print int(W)
print int(S)
print
```

Here is the output after running the introductory section of the program:

211129 Carnot
Liberty Basic

Initial values

P V T Qh Qc W S
100000
24.8
298
0
0
0
0

The second section of the program computes the effect of adiabatic compression of the engine gas sample from ambient to the boiling point:

print "Start adiabatic compression and warming calculation."
while T<373
dV= -1*absdV
V=V+dV
dT=-2*P*dV/R/5
[The previous command is based on U = 5/2 R T.]
T=T+dT
P=R*T/V
dW=P*dV
W=W+dW
wend
print "Output"
print "P V T Qh Qc W S"
print int(P)
print V
print T
print int(Qh)
print int(Qc)
print int(W)
print int(S)
print

Here is the output after calculating the adiabatic compression:

Start adiabatic compression and warming calculation.
Output

P V T Qh Qc W S
219246
14.14
373.061405
0
0
-1560593
0

Comment: Work is done to compress the gas, but no heat flows.

The third section of the program calculates the isothermal expansion of the engine gas sample at 373 K that draws heat Qh < 0 from the high temperature reservoir. Note the choice of the final volume is arbitrary since temperature is constant. Here the mole of gas expands from 14.14 m^3 at the end of adiabatic compression to 45.2 m^3 at the end of isothermal expansion. (This is the step that caused Carnot the difficulty he reported in his memoir.)

```
print "Start isothermal expansion."
while V<45.2
dV=absdV
V=V+dV
P=R*T/V
dQh=-1*P*dV
Qh=Qh+dQh
dW=P*dV
W=W+dW
S=S-dQh/T
wend
print "Output"
print "P V T Qh Qc W S"
print int(P)
print V
print T
print int(Qh)
print int(Qc)
print int(W)
print int(S)
```

print

Here is the output for the isothermal expansion steps:

Start isothermal expansion.
Output
P V T Qh Qc W S
68572
45.21
373.061405
-3602572
0
2041979
9656

Comment: The entropy S is calculated for the engine gas. Qh < 0 represents heat lost by the high temperature heat reservoir. For this part of the Carnot cycle, the heat lost by the hot reservoir appears as work done by the engine gas since the internal energy of the engine gas is constant.

The fourth section of the program calculates the following process: The engine gas sample is expanded adiabatically until its temperature is back to ambient temperature where it began at the start of the cycle:

print "Start adiabatic expansion calculation."
while T>298
dV=absdV
V=V+dV
dT=-2*P*dV/R/5
T= T+dT
P= R*T/V
dW=P*dV
W=W+dW
wend

print "Output"
print "P V T Qh Qc W S"

print int(P)
print V
print T
print int(Qh)
print int(Qc)
print int(W)
print int(S)
print

Here is the output after the adiabatic expansion:

Start adiabatic expansion calculation
Output
P V T Qh Qc W S
31234
79.3
297.98
-3602572
0
3601280
9656

Comment: There is zero change to the engine gas entropy.

The fifth section of the program calculates the isothermal compression back to V=24.8 to complete the cycle of the engine gas sample:

print "Start isothermal compression calculation."
while V>24.8
dV=-1*absdV
V=V+dV
P=R*T/V
dQc=-1*P*dV
Qc=Qc+dQc
dW=P*dV
W=W+dW
dS=-1*dQc/T
S=S+dS

wend

print "Output"
print "P V T Qh Qc W S"
print int(P)
print V
print T
print int(Qh)
print int(Qc)
print int(W)
print int(S)
print

Here is the output after the fifth program step, the isothermal compression that completes the cycle:

Start isothermal compression calculation.
Output
P V T Qh Qc W S
99889
24.79
297.986856
-3602572
2879120
722159
-5

Comment: The total working gas entropy change for the cycle as a whole is small, as expected.

The final section of the program calculates quantities needed to confirm that the first and second laws are satisfied. The final commands are:

print W
print (Qh + Qc)
print (Qh/Qc)
print (373/298)

Here is the output after calculations for the complete cycle have been finished:

722159 J (Cycle work)
-723451 J (Heat consumed)
1.2513 (Heat exchange ratio)
1.2517 (Reservoir temperature ratio)

The final output labels in parentheses are not produced by the computer program but were added afterward. The engine working gas returns to its initial state so that its entropy is restored to its original value. The second pair of results is expected if the engine runs slowly enough so entropy changes can be precisely reversed by running the cycle in the opposite direction. The way Carnot recognized this special capability of a reversible heat engine assures his reputation despite the difficulty with the first law.

Carnot introduced the engine cycle bearing his name as a standard to which the performance of other engines could be compared. For this example, the thermodynamic efficiency is given by

$$(Qh + Qc)/Qh = (-3602572 + 2879120)/-3602572 = .2008$$

or about 20%. Internet sources suggest Watt's breakthrough doubled or tripled the performance of a Newcomen engine while giving a thermodynamic efficiency of only around 3%.

In subsequent chapters we will see that a phase change from liquid to vapor can be used as the basis for approximating a Carnot cycle.

The next step

Even though it is not a useful source of power, the Carnot cycle can provide the basis for a universal approach to measuring temperature. For our Carnot cycle calculation in this chapter, we assumed a large relative expansion at constant temperature with volume increasing by a factor of about three as if we expected the cycle might actually be used in something resembling a traditional steam engine. If we think thermometer in place of engine, there is no reason the change in volume for the isothermal expansion should not be much less than the total cylinder volume and analyzed in a single differential step. (The subsequent adiabatic expansion must still be large enough to cool the engine gas to the temperature of the colder heat reservoir and generally needs to be analyzed in multiple steps as in the computed example.) Then, if we recall the simple argument from Chapter 7 showing the entropy of an ideal gas is a state function, it follows at once that

$$dQh/Th = - dQc/Tc,$$

where the cycle work is the algebraic (appropriately signed) sum

$$dW = dQh + dQc$$

Then it follows from Carnot's early argument (based on the impossibility of a particular form of perpetual motion) that $dQh/Th = - dQc/Tc$ must apply generally to all working fluids suitable for use in heat engines operating in cycles. The idea that cyclets for different substances and the same pair of temperatures must always have the same Carnot efficiency is the foundation of the argument. Rudolph Clausius was responsible for taking it to the next level so it would apply to an arbitrary, reversible path in the pressure-volume plane, as we will see in the next chapter.

(Note that it is possible to work this problem with the heat exiting and entering the heat reservoirs or with the heat alternately entering and exiting the cylinder gas sample. But only the entropy change of the cylinder gas sample is zero for a complete cycle.)

Chapter 11
Clausius' Entropy

About the same time that William Thomson was revising Carnot's work on heat engines operating in cycles to include the first law of thermodynamics, the German physicist Rudolph Clausius was applying the same principles to a more general class of thermodynamic interactions, ones in which a sample exchanged heat at more than just two distinct temperatures. He showed how to extend the idea that ideal gas entropy is a state function (from Chapter 7) and conclude that it is a state function for other substances as well.

Max Planck suggests why Clausius could propose that entropy is a state variable for all substances that can be made to serve in the cylinder of a Carnot engine and not just for ideal gases, as we found it to be by relatively simple analysis in Chapter 7: (Planck, p. 102)

Suppose two Carnot engines, one using an ideal gas G and the other a non-ideal gas G', make reversible heat transfers Q and Q' between a hot reservoir and a cold reservoir. Then, to avoid Carnot's perpetual motion paradox from the previous chapter,

$$Qh/Qc = Q'h/Q'c$$

must follow, and we can use this result to define a universal temperature scale (corresponding to the Kelvin scale) such that

$$Qh/Qc = Q'h/Q'c = -Th/Tc.$$

With temperature defined in this way, we can conclude that

$$Qh/Th = -Qc/Tc,$$

and

$$Q'h/Th = -Q'c/Tc$$

These results follow from the choice of using Carnot engines to define temperature and the impossibility of building composite heat engines that work in cycles without exhausting heat to the environment.

Clausius took this idea a step further:

When heat exchanges are reduced to relatively small increments dQ for the purpose of numerical integration, a closed thermodynamic cycle for an ideal gas sample can be approximated by the sum of contiguous differential Carnot "cyclets" that divide the indicator diagram representing the complete cycle into narrow strips defined by adiabatic expansions and contractions running locally in almost parallel paths across its interior. The construction of these cyclets is easy to visualize if one starts by laying out closely spaced adiabatic curves one after another across an arbitrarily selected closed path for which Sum dQ/T is to be evaluated. Then the short connecting isothermal path segments are dropped in following the outline of the intended path of integration. Positive and negative contributions cancel pairwise, so the total entropy change for the ideal gas sample is zero as it is carried around the loop and back to its starting point in the pressure-volume plane.

This same process can also be applied with the non-deal gas G' and used to divide an arbitrary cycle into a superposition of Carnot cycles. Again, the contributions along the superimposed adiabatic arcs cancel. Then, because these are Carnot cycles, pairs of increments dQh'/T and dQc'/T should also cancel, and the sum of the contributions

$$dS' = dQ'/T$$

should be zero for the whole cycle of the non-ideal gas as well. Otherwise, a cyclet of the non-ideal gas could be singled out and paired with a Carnot cycle of the ideal gas spanning the same reservoir temperatures to form a compound engine that would violate Carnot's principle.

For this argument, note that the adiabatic curves for the non-ideal gas can cut across those for the ideal gas if both sets are plotted on the same diagram.

Clausius would have been aware of the simple argument in Chapter 7 that shows that entropy is a state variable for an ideal gas, and he could have extended the result to include non-ideal gases that do not satisfy $PV = nRT$ by applying this method of cyclets to the non-ideal gas.

M. J. de Oliveira discusses Clausius' approach. (de Oliveira, 2020) But the treatment goes into greater mathematical detail than is possible for us here, for it includes the method of partial differential equations. The form of argument we use, which is based on what might be called a superposition of differential Carnot cycles, has often been presented in textbooks. (Levine, p.83; Sandfort, p. 175; Page, p. 288)

Wayne M. Saslow writes that both William Thomson and Rudolph Clausius had the idea of combining Carnot cycles into a sum representing an arbitrary cycle. (Saslow, 2020)

Entropy measurement

Clausius' form of the second law of thermodynamics can be expressed as

$$dS(sam) + dS(sur) >= 0$$

where *dS(sam)* represents the entropy change of a particular sample of matter as it gains or loses heat and *dS(sur)* represents the entropy change of other matter with which the heat is exchanged.

The way entropy helps explain both the operation of heat engines and also (as we shall see) the existence of well-defined temperatures for phase changes such as freezing and boiling provides strong arguments for accepting the reality of this mysterious thermodynamic quantity. It appears to be a property of all substances and not just ideal gases.

Let us review how the entropy change of a sample might be measured around a closed path in a laboratory. One approach could be to measure entropy change as a function of temperature using a series of water baths at temperatures spanning the range selected. At each step, the heat entering the sample can be found from the specific heat of water and the change in the temperature of the water as it comes into equilibrium with the sample at a new temperature. Variable sample pressure is needed for entropy change to be tracked around a cycle that closes on itself and demonstrates that Sum $dS = 0$ in the limit as the individual steps are made smaller and smaller. Therefore this interesting demonstration seems out of reach for most. (In the Clement-Deshormes experiment from Chapter 7, the entropy change can be tracked even though not all steps in the complete cycle can be observed with the apparatus described there, so a computer simulation is required to close the cycle.)

Entropy values can be expressed in tables, and it is not essential that they be represented by closed form mathematical expressions. Even if an explicit formula for entropy cannot be found, experimental confirmation of the result Sum $dQ/T = 0$ for reversible sample measurements around closed paths in thermodynamic variable space is what gives entropy status as a fundamental physical property of matter.

Chapter 12
Enthalpy Changes in Refrigeration

Throttling the flow of a fluid by gradually closing a valve eventually produces a process characterized by a pressure drop at the constriction and negligible kinetic energy both before and after the valve. The enthalpy H is identified with the state variable $PV + U$, and it does not change as the fluid passes through the constriction since, in the absence of kinetic energy and the flow of heat to or from the throttle housing, the first law requires

$$(PV + U)(upstream) = (PV + U)(downstream)$$

for a mole of fluid passing through. It is assumed that the throttle housing is thermally insulated so that heat conduction does not contribute during steady flow. As a mole of fluid is forced into the throttle by pressure upstream, the molar work is $PV(upstream)$ and the molar internal energy that is brought along is $U(upstream)$. As the fluid emerges, still with negligible kinetic energy, the work done on the fluid downstream is $P\,V(downstream)$, and $U(downstream)$ is the molar internal energy that is carried away.

Consequently, enthalpy

$$H = P\,V + U$$

is a state property that is conserved when a fluid flows slowly through a constriction. This means the temperature of an ideal gas does not change as it passes through a throttle since all the terms in the defining equation are proportional to the temperature.

The situation is more interesting when a volatile liquid is forced through the throttle and partially evaporates. Useful cooling can result, and both the temperature change and the amount that evaporates are functions of the pressure change. The problem can be approached with the help of a thermodynamic table for the substance that gives the specific enthalpies of the liquid and vapor phases as functions of the initial and final pressures *P1* and *P2*. (One can also read the corresponding equilibrium temperatures *T1* and *T2* and the specific entropies if the table is complete.) To determine how much liquid evaporates, one can use the initial value for the liquid phase enthalpy *Hliq* as well as the liquid and vapor phase enthalpies at the final temperature and pressure to solve

$$Hliq(P1) = Xvap * Hvap(P2) + (1-Xvap) * Hliq(P2),$$

for the vapor fraction *Xvap*, the fraction of the sample that goes over to the vapor state at the final pressure. When a liquid in equilibrium with its vapor is treated a mole at a time, pressure *P* and the vapor fraction *Xvap* can serve as thermodynamic state variables. (Later, we will refer to a reference table giving the equilibrium properties of ammonia.)

Flow through a throttle with enthalpy conserved is an important step in most refrigeration systems even though at one point the liquid may spurt vigorously from a nozzle. This kinetic energy is promptly dissipated and goes to heat the refrigerant before it enters the heat exchanger where the bulk of evaporation occurs, so constant enthalpy is a valid approximation at the start of the process and before the significant cooling heat exchange during which the rest of the refrigerant evaporates in the tubes of the heat exchanger.

In practical air conditioners, a liquid refrigerant undergoes a temperature drop and partial evaporation as it flows through a throttle from high pressure to low, and the pressure drop can be controlled (within limits) to give the temperature change desired. There is more going on than when water emerges from a nozzle at the end of a garden hose, and we must be careful to distinguish what is called a throttle or a porous plug from a low-friction nozzle where downstream kinetic energy is significant.

If internal friction in the throttle assembly prevents the refrigerant from gaining appreciable kinetic energy, then an initially liquid refrigerant can emerge as a slowly moving mixture of gas and liquid at a significantly lower temperature. As previously stated, this cannot be the case for an ideal gas like air when there is no change of phase. In this case, equating enthalpy in to enthalpy out requires

$$Pin\ Vin + Uin = Pout\ Vout + Uout,$$

so there can be no temperature change for an ideal gas where the terms are proportional to the Kelvin temperature. This puzzling result (attributed to Gay Lussac and Welter) is noted by Sadi Carnot in a footnote. (Carnot, p. 16) In the limited region in the vicinity of a nozzle where a gas jet is moving rapidly, spectroscopy and schlieren photography can show its reduced temperature, but these optical effects vanish as kinetic energy dissipates. (Skinner, 1980)

When a liquid refrigerant is used, on the other hand, a phase change may occur that satisfies

$$Hliq(P1) = Xvap * Hvap(P2) + (1-Xvap) * Hliq(P2),$$

since the vapor fraction *Xvap* can serve as a state variable in the limited region of thermodynamic variable space where liquid and vapor coexist. The pressure drop from *P1* to *P2* is maintained by a motor-driven compressor, and a temperature change accompanies a redistribution between phases. Since there would be no kinetic energy to give an initial boost, there can be no spontaneous return flow through a throttle from low pressure to high, so throttle flow is irreversible, and entropy increases during the process.

Irreversibility means that the two-phase refrigerator with a simple throttle (or expansion valve) cannot be run backward and generate mechanical power in the way that was possible in the case of the Brayton cycle air conditioner of Chapter 8. However, a two-phase air conditioner can be modified to operate as a (so-called) heat pump to warm houses in winter by drawing heat from cold outdoor air and using adiabatic compression to raise the temperature. Because the coefficient of performance can be significantly greater than unity, there is an opportunity here to reduce dependence on combustion for space heating.

Throttled flow can also be used as a step in liquifying gases. (Sears, p. 306, 330)

There is another important principle that involves the state variable enthalpy. When a sample of substance exchanges heat with its surroundings at constant pressure,

$$dH = dQ$$

So, in chemistry, the heat released in chemical reactions taking place at constant pressure can be measured to determine the specific enthalpies of compounds and their constituents. (Masterton, p. 217) We will consider the role enthalpy plays in determining the equilibrium conditions for phase changes and chemical reactions in the next chapter.

Chapter 13
Phase Changes

In order to discuss two-phase refrigeration systems, we need to consider the entropy and enthalpy changes that accompany a phase transformation, starting this time with the familiar example of melting ice. Similar ideas apply to evaporation from liquid to gas (or vapor), as in the case of water to steam, but ice is familiar and both easier and safer to work with.

Let us consider how we might measure and observe the entropy change that accompanies a phase change at constant pressure. First we imagine the plot of entropy versus temperature that results if we measure the entropy change of freezing with the experimental method given earlier in Chapter 11; that is, by using reference heat reservoirs spanning a range of temperature from a little below to just above freezing.

Each time the ice sample is placed in contact with a thermal reservoir at a slightly higher temperature, the heat dQ gained by the sample can be estimated from the small reservoir temperature change that results, and the sample entropy is incremented by dQ/T. The sample entropy trends upward with temperature until the melting temperature *Tfusion* is reached. Then the temperature remains constant while the entropy continues to be incremented by amounts dQ/T_{fusion} until all the ice is gone and entirely replaced by water. The sum of the dQ during melting equals the latent heat of fusion. The ice and its meltwater must be kept together so the sample consists of two phases with constant total mass until fusion is complete.

The vertical step in the entropy versus temperature plot is where the latent heat of fusion enters the sample from the surround and causes melting. (The corresponding region where liquid water and steam coexist at constant pressure and temperature extends over a much larger relative range of volume, and the work of expansion associated with the phase change is substantial.) Once the ice sample is entirely melted, the temperature resumes its rise and the entropy trends upward at a different slope from before.

During the phase change, two phases of water exist in near equilibrium at a common pressure and corresponding temperature but with distinct molar entropies and enthalpies. If a mole sample of water ice slowly melts at constant temperature and pressure from all solid to all liquid, its molar entropy shifts linearly from S(ice) to S(water) according to the relation

S(solid and liquid together) = X S(water) + (1-X) S(ice), 0 < X < 1.

Here the dimensionless quantity *X* is called the liquid fraction. Introducing the liquid fraction allows us to track the state of a standard quantity of substance transitioning continuously from solid to liquid. (*X* stands in here for the Greek letter chi, which may be used.) The two-phase sample might be a relatively homogeneous slurry of water and ice granules, or it might be an ice cube floating in a cup of water. If the latter, then the ice cube and its melt water should be at least thought of as being kept together in a plastic bag, so the heat that slowly melts the ice is clearly recognizable as coming from outside the ice and melt water sample.

In the limited region of thermodynamic space where the two phases coexist in equilibrium, the liquid fraction *X* (or the vapor fraction) can be treated as an independent state variable when problems are worked with pressure as the other independent variable since, by referring to a table, total molar entropy or enthalpy can be calculated for a mixture of phases at a given pressure. The conditions of equilibrium are relatively straightforward when only two phases of a single compound are present. The situation becomes more complex if two compounds are present. This formidable topic is introduced by Max Plank in his *Treatise on Thermodynamics*. (Planck, 1903, Part IV)

While the ice is melting, the molar enthalpy *H* also shifts linearly from *H(ice)* to *H(water)* according to the same relationship

H(solid and liquid together) = X H(water) + (1-X) H(ice), 0 < X < 1

because enthalpy too is a state function.

What is the temperature T_{eq} at which a sample of water and ice will remain in equilibrium and neither melt nor freeze solid? At constant pressure, the expression that determines the melting temperature T_{eq} for the ice and water is

$$Teq * (Sl - Ss) - (Hl - Hs) = 0,$$

where *Sl* and *Ss* are the liquid and solid molar entropies for water and ice respectively and *Hl* and *Hs* are the corresponding enthalpies. This relationship constrains the melting temperature because at constant pressure, the heat transmitted to the sample from a surround is equal to the change in the sample's state variable enthalpy. If the surround is at the temperature *Teq* chosen to satisfy

$$Teq * (Sl - Ss) \, dX - (Hl - Hs) \, dX = 0,$$

then the surround entropy change will be equal and opposite to the entropy change of the sample when there is a change in the liquid fraction *dX*. This equilibrium can persist indefinitely *since there is no overall entropy change due to either melting or freezing.* The surround must be slightly warmer than Teq for the net entropy gain for sample and surround together to be greater than zero, in which case melting can continue until only water is left. The opposite is true for the case of freezing. This insight is credited to Josiah Willard Gibbs. It also shows the importance of the relationship referred to as Clausius' form of the second law of thermodynamics,

$$dS(sam) + dS(sur) >= 0$$

In other words, if one knows *deltaS(P)* and *deltaH(P)* for the phase change, one knows what the equilibrium temperature *Teq* must be at the given pressure *P*, since there is no net entropy change for either melting or freezing when the sample is placed in a heat bath (surround) where the temperature is

$$Teq = deltaH/deltaS$$

The same argument applies for the change from liquid to vapor. It is based in the idea that both entropy and enthalpy are state variables. Tables have been compiled where liquid and vapor phase entropies *Sliq* and *Svap* and corresponding enthalpy values *Hliq* and *Hvap* are shown by multiple entries that extend over a range of pressures and corresponding equilibrium temperatures. Given access to a suitable thermodynamic table of equilibrium values, perhaps for a refrigerant such as ammonia, it is an interesting exercise to pick a pressure and substitute the respective values from a line of the table into the equation

$$Teq\ (Svap - Sliq) - (Hvap - Hliq) = 0.$$

If one remembers to use the absolute temperature (perhaps in place of a centigrade temperature), the result should come out close to zero, since the same reasoning applies with liquid and vapor phases in place of solid and liquid.

(This familiar explanation for the existence of a well-defined temperature for equilibrium is consistent with observations of how the onset of a phase change can sometimes be delayed beyond the equilibrium temperature. This may be observed when previously boiled water is reheated in a microwave oven, and precautions against being taken by surprise as water bubbles out of its container may be included in the appliance operating instructions.)

To repeat: To keep the experimental condition well defined, ice and its melt water must be thought of as confined in a container and kept separate from the thermal surround to analyze the process of melting (and similarly for evaporation). The container might be a plastic baggie or a calorimeter cup. The thermal surround is essential for this analysis to make sense--the heat cannot come from melt water that makes up the second of two phases!

A familiar experiment is to determine ($Hliq - Hsol$), the heat of fusion for ice, and show it is about 80 calories per gram, where a calorie is the heat to raise the temperature of a cubic centimeter (or a gram) of water 1 degree centigrade. The conversion factor of 4.18 joules to the calorie is the golden thread leading into the wider world of mks units.

Regulating pressure exerts a control on the equilibrium temperature that is relatively easy to detect for ice. The fact that ice is slippery enough to skate on suggests that its freezing point is in fact a function of pressure. Similar dependence on pressure allows a sample of refrigerant with coexisting liquid and vapor phases to be guided around what approaches a Carnot cycle as it produces useful cooling, as we shall see in a later chapter.

Here the melting of ice is used as an example of one phase change among many. The relationship between entropy, enthalpy, and the equilibrium temperature can be extended and applied to chemical reactions. It is important when planning a chemical synthesis to know which side of chemical equilibrium one is on, since no catalyst can trigger a reaction not favored by thermodynamics. (Masterton, p. 702) A tragic drama, the high pressure synthesis of ammonia and the consequences for its discoverer, turned on this. (King, 2012) Geologists see in these same principles the origins of marvelous mineral diversity.

Rubin Battino suggests that the concept of chemical equilibrium is simply too complicated to present in detail in a first year chemistry course. (Battino, 2007) We need to address it here because phase changes are part of practical refrigeration cycles--our air cycle machine in Chapter 8 proved much too bulky to fill the need. In particular, cycles must be analyzed to estimate the COP of a system.

Chapter 14
What Is Entropy?

Rudolph Clausius had the insight to write the second law of thermodynamics as

$$dS(sam) + dS(sur) >= 0$$

where $dS(sam)$ represents the entropy change of a particular sample of matter and $dS(sur)$ represents the entropy change or changes of surrounding matter with which the sample exchanges heat. The law states that when heat passes between samples of material, the total entropy never diminishes. We have already seen this idea at work in the previous chapter. The way entropy helped explain both the performance of heat engines and the existence of well-defined temperatures for phase changes provided strong arguments in favor of introducing this new theoretical quantity.

Clausius postulated that a small amount of heat dQ entering a sample of matter at temperature T raises its entropy S by dQ/T. The change needs to be small and suitable for calculation by the method of finite differences: The result should be almost the same whether the initial sample temperature or the final temperature is used to calculate dS. A compelling result emerges when this process is carefully carried out. The total entropy change of the sample approaches zero when measured around a closed path in thermodynamic variable space. This result shows that entropy is a state variable on equal footing with pressure, volume, temperature, internal energy, and enthalpy.

Unlike the case for internal energy, which could be modeled as the collective kinetic energy of the atoms of an ideal gas, there was no insight at first into the nature of entropy. Here we will see how Ludwig Boltzmann found an atomic model in which entropy would be a state variable. He started from the kinetic theory model of Chapter 7 in which each atom is assigned three independent degrees of freedom x, y, and z and has total kinetic energy

$$Ek = 1/2 \ m \ (Vx^2 + Vy^2 + Vz^2),$$

where m is the atomic mass, Vx is the velocity in the x direction, and so on. $P \ dV$ work was not initially included in his model, which is for constant volume when presented in its simplest form. In this chapter, to follow Boltzmann's argument, we will drop the assumption that we are dealing with atoms a mole at a time. Instead of using the molar gas constant R introduced in Chapter 7 and given by

$$R = Nmol \ k,$$

where $Nmol$ is Avogadro's number, we will introduce No as the total number of atoms and adopt the expression

$$U = 3/2 \ No \ k \ T$$

for *No* monatomic gas atoms. We will see how Boltzmann showed that increasing entropy could represent a potentially spontaneous transformation from a less probable to a more probable distribution of thermal energy among these atoms.

Suppose discrete energy levels are defined to which atoms can be assigned according to their kinetic energies the way individually numbered wooden blocks can be arranged and rearranged on a series of shelves. We do not expect the atoms to all bunch on the same shelf. Experience shows their energies are distributed over a range with varying probabilities. It was plausible to think that Clausius' version of the second law,

$$dSsam + dSsur >= 0,$$

could be based on a natural tendency to proceed from order toward disorder, as when one shuffles a previously arranged deck of cards. A precise rationale for Clausius' expression $dS = dQ/T$ remained for Boltzmann to discover.

The first step turned out to be determining the equilibrium distribution of atoms in energy levels at a particular temperature. The result is known as the Boltzmann distribution. Once it has been found, the level-by-level energy differences between distributions at two slightly different temperatures can be added to calculate first the quantity

$$dQ(dT)$$

and then

$$dS(dT) = dQ(dT)/T$$

after division by the average of the two Kelvin temperatures. When calculated in this way, *dS(dT)* turns out to have the properties of a state variable.

What is W?

In this chapter, we have run through our familiar variable names and will use W, previously the symbol for work, in a new and unfamiliar way: To follow Boltzmann's reasoning, we must count the different ways W that a group of atoms can be distributed among energy levels. (Leighton, Ch. 10)

Let us begin by considering the relative probabilities of obtaining different results when rolling two dice. The likelihood of a given number coming up is determined by counting the number of ways it can occur when the dice are rolled. Based on the expectation that the dice are fair, one should bet on rolling the total that can come up in the greatest number of different ways, which is seven.

To continue with a second example, imagine a long list showing all the permutations of the way N individually numbered items can be laid out in a row. For $N =$ nine and with the items identified by the numerals one through nine, the start of the list might look like

$$123456789$$
$$213456789$$
$$231456789$$
$$234156789$$
$$234516789$$
$$234561789$$

and so on for pages and pages to show all the W possible arrangements. There is no apparent pattern to follow the way there is for a mechanical counter with drums in a row representing the different powers of ten. However, the total number of permutations is straightforward to calculate because there are nine ways to choose the the first item, 8 ways to choose the second, and so on. More than 300,000 rows are needed here. All orderings are considered distinct at this point in the discussion, but that assumption will now be modified:

For Boltzmann's problem, the items become atoms, and the axis along which the atoms are placed is divided into discrete energy bins extending from zero to as high an energy as turns out to be needed once the average fraction of the atoms that is in each energy interval can be determined. This equilibrium distribution will vary in a predictable way with temperature.

To discover the pattern of the equilibrium arrangement, we need to consider the number of ways W that No atoms can be distributed among energy divisions that are designated by the index $i = 1, 2, 3$, and so on. But this number is no longer the one in the last example; that is, not what someone would obtain if they counted all the ways of laying the No atoms in a row. In Boltzmann's approach, the number is substantially diminished, for he counted as one all the different orderings of the same Ni particles in the *ith* energy cell.

For Boltzmann's distribution, No is the total number of atoms, and $No!$ is the total number of arrangements or orderings without energy divisions, since there are No ways to choose the first atom, $No-1$ ways to choose the second, and so on. (Recall that $3! = 3 * 2 * 1 = 6$, and so on.) Boltzmann postulated that inserting energy divisions diminished the total number of arrangements because different orderings of atoms with the same energy were indistinguishable. The precise way the partitions reduce the total number of arrangements can be shown as follows:

Suppose W is the number of ways of arranging No particles in energy levels with $N1$ particles in the first cell, $N2$ particles in the second, and so on until all No particles have been assigned. Also, suppose the different possible orders of particles in a level are indistinguishable. Next, consider what happens to the number of arrangements as the requirement of indistinguishability is removed level by level. Each time the indistinguishability requirement is dropped, say for the *ith* energy level, the number of distinct arrangements or permutations is increased by a factor of $Ni!$. Now carry this process through until all levels have been treated in this way. Then, because we saw that there are $No!$ possible orderings of No particles, it appears that the number of permutations is given by

$$W = No!/\text{Product}(i=1 \text{ to } n)Ni!$$

even though there may be no easy way of laying counters out on a table and systematically testing the conclusion.

To continue, it is convenient to introduce lnW, the natural logarithm of W, which we can assume to be an integer much greater than zero. The signature property of the logarithm function is that the logarithm of the product of two numbers is equal to the sum of the logarithms of the numbers. This property follows if the value of the logarithm of a number x is found by summing $dlnx = dx/x$ from $lnx(x = 1) = 0$ to the desired value of x, which may be represented by X. Then, to show generally that $ln(x\,y) = ln(x) + ln(y)$, we note that $dln(x\,y) = (x\,dy + y\,dx)/(x\,y) = dx/x + dy/y$, so the logarithm of a product $X\,Y$ can be found by the integration of $dx/x + dy/y$ from $x, y = 1, 1$ to $x, y = X, 1$ and then to $x, y = X, Y$, which gives the sum $ln(X) + ln(Y)$. Notice how the argument we used in Chapter 7 to show that the entropy of an ideal gas is a state function applies here as well.

Boltzmann's distribution

Next we use the natural logarithm function to re-express the equation for the number of permutations as

$$lnW = ln(No!) - \text{Sum}(i=1 \text{ to } n)ln(Ni!)$$

Now it is possible to estimate the change in lnW caused by a change dNi in the number of atoms in the *ith* energy level by using the approximation

$$dlnW(Ni, dNi) = -ln(Ni)\,dNi$$

Then, for simultaneous small population changes among all the levels, the approximate change in the logarithm of the number of permutations is

$$dlnW = -\text{Sum}(i=1 \text{ to } n)\,lnNi\,dNi$$

To continue, we introduce another idea, that natural fluctuations dNi in the numbers of atoms in different levels must conserve internal energy U when a gas is in thermal equilibrium and insulated from heat exchanges so

$$dU = \text{Sum}(i=1 \text{ to } n)\ Ei\ dNi = 0,$$

where U is the internal energy and Ei is the energy of individual atoms in level i.

To combine these, we assume the energy bins are of equal width and introduce the convention that $Ei = i\ Eo$, where i is the integer representing the ith energy level. Then for fluctuations with energy conserved, the last equation becomes

$$dU = \text{Sum}(i=1 \text{ to } n)\ i\ Eo\ dNi = 0$$

Boltzmann could compare this with the previous result for the change in the logarithm of the permutations,

$$dlnW = -\ \text{Sum}(i=1 \text{ to } n)\ lnNi\ dNi,$$

and conclude that a distribution of atoms among energy states of the form

$$lnNi = \text{first constant} + \text{second constant} * i*Eo$$

assures that lnW also fluctuates in the vicinity of a steady value.

To continue, the expression for Boltzmann's distribution of atoms over energy states can be written out as:

$$lnNi = Lo - i*Eo/Eref,$$

where Lo is a constant leading term chosen so that summing the numbers of atoms distributed among the energy states gives the total number of atoms No that are in the sample. It turns out the choice of reference energy $Eref$ we want is

$$Eref = 3/2\ k\ T,$$

where Boltzmann's constant k and Avogadro's number No are related to the specific heat at constant volume Cv by

$$Cv = 3/2 \, No \, k$$

Then the Boltzmann distribution of atoms over kinetic energy levels becomes

$$lnNi = Lo(T) - i*Eo/(3/2 \, k \, T)$$

Now we have found a law describing how the distribution of atoms over energy states depends on temperature. There must be many atoms and Eo must be small compared to $3/2 \, k \, T$ for the distribution of atoms among energy states to assume its characteristic form with the most atoms in the level of lowest energy. Then the numbers Ni diminish smoothly toward higher energies.

Knowledge of this law allows us to identify entropy S with the state variable $3/2 \, k \, lnW$, and we can show that Clausius' result

$$dS = dQ/T$$

holds when Boltzmann's theoretical model is used to calculate an entropy change.

The state variable lnW and Boltzmann's entropy

We can now calculate how the equilibrium distribution of gas atoms among energy levels changes with temperature, and from equilibrium distributions at two slightly different temperatures, we can also calculate the accompanying change in the logarithm of W, the number of equivalent permutations of the atoms distributed in the levels. This is done by combining the previous results,

$$lnNi = Lo(T) - i*Eo/(3/2 \, k \, T)$$

and

$$dlnW = - \text{Sum}(i=1 \text{ to } n) \, lnNi \, dNi$$

The term in Lo drops out because the sum of the dNi is zero, so this gives

$$dlnW = - \text{Sum}(i=1 \text{ to } n) \, (- i*Eo/(3/2 \, k \, T)) \, dNi$$

or

$$3/2 \, k \, dlnW = \text{Sum}(i=1 \text{ to } n) \, (i*Eo/T) \, dNi = dQ/T$$

Therefore, it appears that the entropy of the atoms in a monatomic gas sample corresponds to the statistical quantity $3/2 \, k \, lnW$ to within an additive constant.

We can conclude that spontaneous flow of heat contributes directly to the increase in atomic scale disorder W according to

$$3/2 \, k \, dlnW = dQ/T,$$

even though we may have little curiosity about the enormous quantity W itself. The approach to calculating $dlnW$ from two statistical distributions for slightly different temperatures is demonstrated by the program given below, and the result is checked by also calculating dQ/T using the coefficient of heat at constant volume Cv we found in Chapter 7,

$$dQ(dT) = Cv \, dT = 3/2 \, No \, k \, dT,$$

and substituting the value in Clausius' expression

$$dS = dQ(dT)/T$$

So, to within an additive constant, $3/2 \, k \, lnW$ appears to be the physical property of a monatomic gas sample whose change we are tracking by measuring a small isochoric entropy change in the laboratory. (Born, Ch. 1.6) In the next section, we will check this result by means of a numerical simulation. Then we will conclude the chapter by showing a way to include the effect of an adiabatic expansion in the calculation.

A numerical simulation

We have used a computer program to check the method for evaluating entropy change suggested in this chapter. In the listed program, dS is calculated from differences between two distributions for slightly different temperatures, and then the value is checked by comparison to dS(dT) found as

$$dS = C_v \, dT/T = 3/2 \, N_o \, k \, dT/T,$$

where Cv is the coefficient of heat at constant volume for a monatomic ideal gas.

Here No is the total number of atoms in the sample. 1000 atoms are used in the simulation, and the temperature change between the distributions is one part in a hundred. The cells of lowest energy are occupied by about 95 atoms, and both distributions roll off close to zero before the program index i reaches 100. (It is assumed numerical averaging models the role fluctuations would play in an actual experiment.) The entropy change by Boltzmann's method is 14.923 entropy units, while evaluating 3/2 No k (Th-T)/T gives a change of 15 units.

The program dS(dT) 220426 sw calculates the entropy change Sb for warming at constant volume by Boltzmann's approach. The values for the normalizing constants Lo and Loh were determined at the computer by trial and error.

Command list:

```
print "dS(dT) 220426 sw"
k=1
No=1000
T=100
Th=101
```

Eo=3*k*T/20

E=0
Eh=0
N=0
Nh=0
deltaS=0
deltalnW=0
dN=0
dSb=0
Sb=0
U=0

Lo=4.65564
Loh=4.64519
Eref=3*k*T/2
Erefh=3*k*Th/2
print "Eref = "; Eref
print "Erefh = "; Erefh

print
i=0
while i<100
i=i+1
lnNi=Lo-i*Eo/Eref
lnNih=Loh-i*Eo/Erefh
Ni=exp(lnNi)
N=N+Ni
Nih=exp(lnNih)
Nh=Nh+Nih
U=U+Ni*i*Eo

dNi=Nih-Ni
dSb=-1*3*k*lnNi*dNi/2
Sb=Sb+dSb

[Sb represents the entropy change between the two distributions calculated by this program.]

'print i, int(Ni) ,int(1000*dNi)/1000, U

wend

print
print "Output:"
print
print "Sb = "; Sb
print "3/2 No k (Th-T)/T = "; 3*No*k*(Th-T)/T/2

print
print "Lo = "; Lo
print "Loh = "; Loh
print "N = "; N
print "Nh = "; Nh
print
end

Program Output:

dS(dT) 220426 sw

Eref = 150
Erefh = 151.5

Output:

Sb = 14.922991
3/2 No k (Th-T)/T = 15

Lo = 4.65564
Loh = 4.64519
N = 1000.00778
Nh = 1000.00664

Boltzmann's thermodynamic cycle

The situation becomes more complicated if the volume is not held constant and entropy change is calculated along an arbitrary path in thermodynamic variable space that can be represented by adiabatic volume increments and isochoric temperature increments in alternation. How can we extend Boltzmann's statistical model to calculate entropy changes generated in this way and trace out a complete thermodynamic cycle?

For the isochoric steps, we have found that lnW changes so

$$3/2 \, k \, dlnW = Sum(iEo \, dNi)/T = dQ/T$$

To calculate entropy change along a two-dimensional path, the height of an isochoric temperature step can be incremented by the effect of a preceding adiabatic volume step as shown on the cover of this ebook. The thermodynamic expression for $dS(dT)$,

$$dS(dT) = No \, k \, 3/2 \, dT/T$$

now becomes

$$dS(dV, dT) = No \, k \, (3/2 \, dT/T + dV/V),$$

where the final term in dV/V takes into account the small relative volume change due to an adiabatic step prior to the isochoric step, as suggested on the cover of this ebook.

The adiabatic volume changes are incorporated into the model by assuming that the energy cell occupancy numbers remain unchanged while Eo and T expand and contract simultaneously so the ratio Eo/T remains the same: this assures that lnW will not change during the adiabatic temperature change produced by the adiabatic volume change dV. In this model, T and consequently Eo return to their initial values at the end of a cycle, so lnW will do the same, and a full calculation in which the numerical model traces out a cycle should confirm this. Here, however, these conclusions have been checked numerically only for a single isochoric step with its temperature change. The program above only compares the sum of terms $iEo \, dNi(dT)$ to the change in internal energy given by $3/2 \, No \, k$

dT. No attempt has been made to extend the method to explicitly include adiabatic temperature shifts and fully demonstrate the computer model for a complete thermodynamic cycle.

Boltzmann's ideas met with initial resistance. Even well into the nineteenth century, the idea of atoms remained controversial despite the successful use of the law of multiple proportions for explaining chemical reactions. Later, however, Boltzmann's statistical approach was successfully adapted to treat blackbody radiation by appealing to new quantum mechanical principles. (Planck, *Heat Radiation*, Part III)

Chapter 15
Cooling with Nozzles and Throttles

Now we are prepared to consider the approach used in practical air conditioners where cooling is initiated by allowing liquid refrigerant to flow from high pressure to low and into the coils of an indoor heat exchanger, which is often located in the plenum of an air duct system. There is an initial temperature drop, and cooling is produced as the refrigerant boils at the reduced pressure and corresponding boiling temperature maintained by the action of the compressor, which acts as a suction pump on its intake side. Heat is removed from the air circulating past the cooling coils by a process resembling water evaporating on a hot stove top that takes place inside the tubes. The compressor functions similarly to the Brayton cycle compressor of Chapter 8, and the initial kinetic energy of the refrigerant spurting into the cooling coil turns out to be almost negligible compared to the heat of evaporation that accounts for most of the cooling effect.

In the Brayton air conditioner, all the cooling effect came from expanding the air adiabatically in an expansion engine, and there was no cooling contribution from a phase change. In practical, two-phase air conditioning, the situation is very much reversed. The possible cooling benefit associated with adiabatic expansion can be small compared to cooling by evaporation, so the equivalent of the expansion engine is not missed.

Nevertheless, a nozzle used in combination with a single-stage impulse turbine and dynamo configured to recover kinetic energy is a practical way to achieve the highest possible efficiency from a two-phase refrigeration system. The kinetic energy is removed as useful work by the turbine before it can be converted to heat through friction. However, this approach is unusual apparently because the gain from the added complexity is small. (Brasz, 2003) Thus in a conventional, two-phase refrigeration system, kinetic energy is promptly dissipated by friction after a liquid refrigerant spurts through a nozzle and separates into a vapor fraction and a liquid fraction at reduced temperature and pressure. Relatively efficient cooling still results as the remaining fluid evaporates while passing through the heat exchanger. For home cooling, the friction heating by the moving jet turns out to be a minor loss rather than a major one.

Refrigeration with a throttle and two-phase flow

Unlike the case with an ideal gas, useful refrigeration can be produced when phase separation is produced by a throttle. To examine this process, we assume tables are available giving equilibrium temperature T, liquid phase enthalpy *Hliq*, gas phase enthalpy *Hvap*, liquid phase entropy *Sliq*, and gas phase entropy *Svap* as functions of pressure for the refrigerant. Here we would expect the enthalpy and entropy values to be given on a per mole basis. Instead, we will work examples using a table where these quantities are given on a per kilogram basis. This is not a disadvantage as long as we restrict our attention to a single refrigerant compound.

In a typical air conditioning system, refrigerant is condensed at a temperature *Th* somewhat above the outdoor ambient temperature and flows as a liquid to the indoor heat exchanger. The liquid spurts through a nozzle into the entrance of the indoor heat exchanger coils and cools to the temperature *Tc* corresponding to the vapor pressure maintained by the compressor, which acts as a vacuum pump for gas entering its intake port. The kinetic energy of the fluid exiting the nozzle is converted to heat by friction at the entrance to the cooling coil, which constitutes a small loss to available cooling by the system overall. The model of constant enthalpy (rather than constant entropy) applies to this step in the process.

Most of the refrigerant that enters the cooling coil in this way is still in the liquid state at the temperature *Tc* corresponding to the vapor pressure maintained by the compressor. All of this liquid can evaporate in the cooling coil at close to constant pressure, and the corresponding latent heat of vaporization is available to cool air from the house interior that is circulating past the outside of the coil at a slightly higher temperature.

Because enthalpy is conserved in a throttling process, the relative amount of liquid refrigerant that is left to cool the interior of the house can be found by refering to a thermodynamic table for the refrigerant being used and solving

$$Hliq(Th) = X\,Hvap(Tc) + (1-X)\,Hliq(Tc),$$

for *X*, which in this example is the fraction that evaporates before the refrigerant enters the heat exchanger tubes. Introducing the vapor fraction as a variable allows us to find the amount of liquid remaining down stream from the throttle. Useful cooling is produced when this liquid evaporates in the indoor heat exchanger.

We do the calculation using values from a thermodynamic table for ammonia. The initial enthalpy at *Th* = 40 C is the liquid enthalpy *Hliq* read from the appropriate column of the table. To find *X*, the vapor fraction after the ammonia is throttled to the vapor pressure at *Tc* = 0 C, we assume enthalpy is conserved, so

$$Hliq(Th) = (1-X)\,Hliq(Tc) + X\,Hvap(Tc)$$

Rearranging gives

$$X = [Hliq(Th) - Hliq(Tc)]/[Hvap(Tc) - Hliq(Tc)]$$

Filling in the quantities from the table with $Th = 40$ C and $Tc = 0$ C gives

$$X = (533.79 - 343.15)/(1605.40 - 343.15) = .151032,$$

so 15.1% of the liquid is lost to refrigeration. This can be compared to the result when work is recovered with the help of a turbine so that the entropy is constant rather than the enthalpy.

Ammonia data generously posted on the Internet by BOC, a supplier of refrigerants, was used for this calculation, which was subsequently checked with data published by Haar and Gallagher. (Haar, 1978)

Improvement with a turbine and constant entropy

If kinetic energy is promptly recovered by means of a spinning turbine wheel as refrigerant emerges from a nozzle into the indoor heat exchanger, then entropy is conserved instead of enthalpy, and very little kinetic energy is converted to heat before liquid refrigerant enters the heat exchanger tubes. (Hays, 1996) This translates to a small improvement in the amount of cooling the system can produce starting from the same amount of electrical energy used to run the compressor and fans. To demonstrate this, we can use the same thermodynamic properties table as before. The calculation process is similar, but we use the entropy data rather than the enthalpy data:

Again, let temperature be the independent variable in the table we are using. Let liquid refrigerant cool as it expands from all liquid at $Th = 40$ C and the corresponding vapor pressure $P(Th)$ to a fast moving mixture of gas and liquid at a lower temperature $Tc = 0$ C and the corresponding vapor pressure $P(Tc)$. The mixture of gas and liquid is brought almost to rest as it reversibly transfers its kinetic energy to the turbine wheel and shaft. To estimate the amount of liquid still available for evaporative cooling, we will solve for X in the same way as for enthalpy using tabulated values for entropy in place of the values for enthalpy. The equation becomes

$$Sliq(Th) = X\,Svap(Tc) + (1-X)\,Sliq(Tc)$$

for the approach where a turbine driving a dynamo used for power recovery. The initial entropy at 40 C is the liquid entropy read from the appropriate column of the table. Then

$$X = [Sliq(Th) - Sliq(Tc)]/[Svap(Tc) - Sliq(Tc)]$$

Filling in the corresponding values from the table gives

$$X = (2.1161 - 1.4716)/(6.0926 - 1.4716) = .13947$$

Since 13.95% of the ammonia evaporates during the reversible, constant entropy expansion (compared to 15.1% for the throttling process), slightly less latent heat of evaporation is lost to refrigeration.

Ammonia as a refrigerant

Unlike the ether Perkins used, ammonia is a practical refrigerant that is still employed. Its applicability for refrigeration purposes suggests itself naturally, since water will condense spontaneously and turn to frost on the outside of a pressure bottle containing liquid ammonia if it is set upright and vented so the vapor that boils off can escape into the air. (This is usually not a good thing to do.)

Ammonia is used as a refrigerant despite another disadvantage: In addition to being toxic, it quickly corrodes copper. Thus iron may be substituted for copper and brass in refrigeration machinery with ammonia refrigerant. A configuration in use decades ago employed a shaft seal made of graphite so an electric motor running in air could drive the compressor. The seal configuration resembled what is used in some food blenders, and a cylindrical metal bellows concentric to the shaft maintained the contact pressure needed.

Applying the Carnot cycle approximation

Now we are almost ready to analyze the performance of ammonia in a complete refrigeration cycle. We are going to apply the Carnot cycle as an approximation, but first we will review the changes to the standard refrigeration configuration that that are needed to make the approximation apply more precisely even though neither change is required to produce a satisfactory air conditioner.

Of course, we will assume that a kinetic energy recovery turbine is used with the expansion nozzle so constant entropy applies rather than constant enthalpy. We can also imagine a hypothetical configuration employing a venturi atomizer in which a mist of refrigerant droplets at temperature Tc is injected into the vapor at the start of compression so the refrigerant is adiabatically compressed to the condensation temperature and pressure instead of being adiabatically superheated and then cooled isobarically before condensation can begin. The reverse of this process occurs in a steam engine when the intake valve shuts off steam coming from the boiler and the expansive part of the power stroke commences. This accounts for the clouds of condensed vapor familiar from old movie footage showing steam locomotives in action. This near-adiabatic process has been described by William Ripper in a textbook on steam engines. (Ripper, Chapters 1 - 3) The difficulty of achieving so-called wet (versus dry) compression is acknowledged by Wilbert F. Stoecker in the *Industrial Refrigeration Handbook*. (Stoecker, p. 52)

For our analysis of this process, we will want to substitute different thermodynamic coordinates for pressure and volume. The choice of temperature and entropy is convenient, and the vapor fraction variable is needed as well to take phase change into account.

There is an additional obstacle to achieving performance economy goals. Air conditioning can be very welcome on a muggy day when an outside temperature "in the nineties" is being reported on the news. On a hot summer day in the writer's community, however, refrigerant condensation at 115 F is plausible, and the coefficient of performance (COP) must suffer as air conditioning becomes more and more desirable. As P N Ananthanarayanan puts it:

"It is clear that by increasing the evaporating temperature and decreasing the condensing temperature, the capacity and efficiency of a vapor-compression refrigeration system improves. Therefore, it follows that we should always strive to design and operate the refrigeration system at the highest possible evaporating temperature and lowest possible condensing temperature to get the best out of the system." (Ananthanarayanan, p. 52)

Chapter 16
Two Phase Refrigeration Cycle

In the Brayton cycle air conditioner example in Chapter 8, we assumed an open cycle where cold air exhausting from the expansion engine circulated through the house to provide cooling. Another approach would be to provide an indoor heat exchanger that would transfer heat from air inside the house to cold air from the expansion engine without allowing them to mix. This would certainly produce a quieter house, and the potential problem of lubricant contaminating the conditioned air would be eliminated. Also, the gas circulating in the heat exchangers could be at much higher than ambient pressures, and this might make the system more compact. A desiccant could dry the air circulating in the loop and eliminate the problem of internal frosting.

The closed loop with two heat exchangers, one hot and the other cold, turns out to be the method of choice, even though there must be additional heat transfer loss associated with the use of a second heat exchanger. However, instead of using an ideal gas as the working fluid, a two-phase approach is better. The refrigerant cycles back and forth between a dense, incompressible liquid state and a gas state that behaves differently from an ideal gas because it is so close to condensation. (Steam from a teakettle was mentioned earlier as an example of this condition.) The expansion engine is replaced by a throttle, and the pressure drop at the throttle determines the equilibrium temperature for the two phases in contact passing through the indoor heat exchanger. Note that this boiling temperature must be somewhat lower than the temperature of the air emerging from the cooling ducts.

It is practical and useful to plot thermodynamic data for this process with enthalpy and pressure displayed on the principal axes. (Ananthanarayanan, Ch. 6) Here we will use entropy and temperature so the process can be analyzed as an approximation to a Carnot cycle. (Sandfort, p. 213)

Let us begin with a simplified overview of the two-phase refrigeration cycle. When the air conditioner operates, liquid refrigerant flows to the evaporator unit located inside the air conditioning duct plenum, which may be located in the attic space of the house several meters or more from the compressor-condenser unit.

Crucially, the liquid flows through a nozzle and picks up speed as it enters the header section of the evaporator coils, where the pressure is greatly reduced by the compressor acting as a vacuum pump at the far end of the refrigerant return line. Bernoulli's principle applies here if the onset of boiling is delayed until after the refrigerant clears the nozzle One reference suggests this coincides with the jet breaking up into a spray of droplets. (Reitz, 1990) It is plausible that the throttling process is completed by friction as the refrigerant sprays into the heat exchanger header. Then, while liquid refrigerant remains, its condition is similar to that of drops of water sizzling in a heated frying pan: The refrigerant continues to boil and absorbs heat at the lower temperature until only vapor emerges from the evaporative heat exchanger and starts on its way back to the compressor to repeat the process. The boiling is sustained by warm duct air blowing past the tubes of the indoor heat exchanger: this is the point at which the cooling felt by the inhabitants of the house takes place. The friction loss as the liquid jet slows down is small compared to the heat absorbed by evaporation as the remaining liquid passes through the heat exchanger and then enters the return line to the compressor as vapor.

Let us look more closely at other aspects of this cycle the way we did with the Brayton cycle. The motor-powered compressor is typically housed with the condensing heat exchanger as a single unit located outside the house. Vaporized refrigerant coming from the evaporating coil in the plenum of the air duct system can be at close to room temperature. It is heated adiabatically in the compressor cylinder, where its pressure is raised to the pressure at which refrigerant is condensing to liquid in the condenser coil, which is kept at near the outdoor ambient temperature by a fan. As the pressures equalize, a pressure-actuated exhaust valve opens, and compressed refrigerant vapor flows from the compressor cylinder into the cooling coil during the remainder of the stroke, very much as air did for our Brayton cycle example.

The gas coming from the compressor is usually superheated; that is, no longer in equilibrium with liquid mist, and it must cool down at constant pressure at the start of its journey through the condenser before it can begin to liquify. This complicates the thermodynamic cycle and makes it depart from the Carnot ideal, since heat is lost over a range of temperatures before liquid starts to form.

Because heat is being exchanged over a range of temperatures, there is a problem at this point similar to the one Carnot had with Watt's steam engine, the one which led him to think about an engine working with an ideal gas. Here, as with the steam engine, practical advantage outweighs theoretical concerns, and this loss is accepted. It is possible the difficulty might be bypassed by spraying a small amount of liquid refrigerant into the compressor cylinder at the start of the compression stroke, perhaps with a device resembling the carburetors formerly used with automobile engines. Evaporation of these droplets could prevent superheating during the compression stroke. We will assume this can be accomplished for calculations later in the chapter even though there are difficulties associated with this approach. (Ananthanarayanan, p. 162)

The temperature at which vapor condenses in the outdoor heat exchanger must be significantly above the temperature of outdoor air so heat can be removed by conduction. Since the compressor cylinder head valves open automatically as pressures equalize, there is the possibility that refrigerant might circulate continuously without ever condensing. This possibility might be eliminated by means of a float valve in the liquid refrigerant reservoir that is adjusted to open at a predetermined depth of liquid refrigerant.

The compressor and electric motor are often sealed inside a common housing. This reduces the problem of refrigerant leakage compared to the alternative where a shaft seal is used between the electric motor and the compressor.

Superficially, the compressor may resemble a small, four-stroke gasoline engine, but the operating conditions are different, and pressure-operated valves eliminate the need for a cam shaft. Lubrication is a concern because the lubricant circulates in limited amounts through the entire refrigeration loop. It must reach the parts of the compressor that require lubrication and, like the liquid refrigerant, never accumulate in the cylinder head space. (Ananthanarayanan, Ch. 12)

There are interesting alternatives to the reciprocating compressor, but these will not be discussed here.

Automatic controls are needed in addition to a thermostat. If the temperature in the return line to the compressor becomes too cold, the flow of liquid refrigerant into the expansion nozzle is cut back by a thermostat-expansion valve so liquid refrigerant will not evaporate in the return line. (Ananthanarayanan, p.104)

In some low-capacity refrigeration systems, the expansion nozzle and thermostatic expansion valve can be replaced by a coil of capillary tube. Only on/off control by a thermostat is required, and there is a beneficial reduction in parts count. (Ananthanarayanan, p. 101)

Controlling a two-phase air conditioner

Suppose an air conditioner is cooling a house on a hot day. The interior temperature falls slowly until it reaches the thermostat set temperature and power to the compressor and heat exchanger fans is automatically switched off. What will happen as the house temperature begins to rise, and how will steady state operation be re-established when the temperature reaches the second thermostat set point and the power comes on again?

Special design features must assure the air conditioner can start automatically when the thermostat switches on the compressor and the heat exchanger fans:

It is assumed that a thermal control valve cuts back the amount of liquid refrigerant entering the indoor heat exchanger whenever the temperature at the entrance to the refrigerant return line falls too low. Therefore, a small amount of entrained lubricant should be the only liquid reaching the compressor intake. Separation of this lubricant can take place in the compressor crankcase while the refrigerant vapor passes through to enter the compressor cylinder intake with a minimum of liquid entrainment. Wiper rings on the piston can also be used to limit oil intrusion into the head space of the compressor cylinder. The compatibility of refrigerant and lubricant must be considered, since they cannot be entirely prevented from coming into contact. (Ananthanarayanan, p. 70)

Liquid refrigerant condensed in the outdoor heat exchanger may be held in a reservoir at near outdoor ambient temperature. During shutdown, pressure will equalize around the system, and a standby heater might be provided so refrigerant vapor cannot condense in the compressor. A float valve can be used to assist startup after a shutdown. Configuring the float to hold the valve clear of its seat subsequent to opening can assure that the line to the throttle will fill with liquid before the it closes again.

The AC compressor motor start-up capacitor may fail without warning and shut the system down. This problem can be easy to fix, but extreme caution is called for because of the danger of electric shock.

Example: A two-phase Carnot cycle

A two-phase refrigeration system might follow a Carnot cycle closely if two modifications could be employed: First, a mist of refrigerant droplets would be sprayed into refrigerant vapor entering the compressor cylinder to eliminate superheating. Second, a throttle loss recovery turbine would be used to extract useful work from the liquid refrigerant as it sprayed into the indoor heat exchanger header.

These refinements have proven unnecessary in practice, but their possibility encourages analyzing the air conditioning refrigeration process as a Carnot cycle. We will not be far from reported performance because so much heat is exchanged at constant pressure and temperature during the condensation and evaporation stages of the practical refrigeration cycle as it is commonly employed. We apply the Carnot cycle model even though superheating in the compression part of the cycle is accepted in practical air conditioning systems as well as some kinetic energy loss by friction during the nozzle expansion so the expansion is isenthalpic rather than more nearly isentropic and reversible.

As already suggested, the temperature differences needed for the indoor and outdoor heat exchangers to function represent significant additional departures from the ideal Carnot cycle.

For our example, we will use ammonia as the refrigerant. To perform quantitative analysis, we use the vapor fraction X to keep track of what portion of a kilogram sample of ammonia is gas and what portion is liquid. The approach follows the Chapter 13 discussion of the melting of ice, except that the liquid-vapor phase transition is involved, not the solid-liquid transition. We will continue to calculate thermodynamic effects as a fraction of a standard quantity of refrigerant changes from liquid to vapor and back, and we will again encounter the kilogram of mass as the standard quantity of ammonia in our thermodynamic table rather than the kilogram mole.

In our analysis, we take advantage of the fact that the Carnot cycle is a rectangle when plotted on a temperature versus entropy chart. (Ripper, Chapter III)

If we refer to a temperature versus entropy diagram for a two-phase Carnot refrigeration cycle, we see that a mole of refrigerant can exist anywhere between 100% liquid and 100% vapor at T_h, the hotter, condensing temperature for the cycle. Thus the total per kilogram heat of evaporation can be used to calculate the entropy change dQ_h/T_h as refrigerant condenses in the outdoor heat exchanger.

At the cooler evaporating temperature T_c where heat is absorbed from house interior air, on the other hand, the vapor fraction X varies from around 10% where the refrigerant emerges from the expansion nozzle to around 90% at the start of compression, where most (but not quite all) of the refrigerant has evaporated to provide cooling. This corresponds to the situation when the cycle is a perfect rectangle in a temperature versus entropy graph, and not to the practical form of the cycle.

In our example, we will see that about 15% of a kilogram of refrigerant will either evaporate during adiabatic expansion or condense during adiabatic compression, so only about 85% of the per kilogram heat of evaporation at 273 K will apply toward calculating the entropy change dQ_c/T_c.

This reduction follows because the range of entropy over which liquid and vapor are in equilibrium increases as the temperature diminishes, as can be seen from inspecting plots of temperature versus entropy at different vapor pressures.

We will use tabulated data for ammonia refrigerant to calculate the work that must be supplied for a standard quantity of refrigerant to pass around the cycle.

For our example, inspection of the thermodynamic data for ammonia shows that the efficiency of a rectangular Carnot cycle can be calculated for ammonia refrigerant undergoing first an adiabatic expansion from all liquid to 9% vapor and then an adiabatic compression from 94% vapor to 100% vapor at the inlet to the outdoor heat exchanger. In this example, the corresponding temperatures are 298 K for condensation and 273 K for evaporation. Between the extreme entropy values, the vapor fraction X changes at constant temperature as the refrigerant absorbs or loses heat in heat exchangers. In a very noticeable change from usual practice, not all the refrigerant evaporates in the cooling coil: A mist of droplets of liquid that have not yet evaporated at the start of the compression stroke is relied on to eliminate the superheating that is a readily visible feature on a plot of the refrigeration cycle as it is usually carried out and displayed in graphs.

Returning to our example, the hotter condensation temperature Th is 25 C or 298 K and the cooler evaporation temperature Tc is 0 C or 273 K. To analyze the cycle, we start at the upper left-hand corner of the rectangle representing the Carnot cycle as it is usually displayed:

As a kilogram of 100% liquid ammonia refrigerant begins adiabatic expansion at 298 K, its liquid entropy $Sliq$ is read from the BOC chart as 1.8804 kJ kg^-1 K^-1. As it expands, its entropy remains constant assuming that kinetic energy is removed as work by a throttle loss recovery turbine. To accomodate this, it separates into liquid and vapor fractions $Xliq$ and $Xvap$ that satisfy

*Xliq(273 K) * Sliq(273 K) + Xvap(273 K) * Svap(273K) = Sliq(298K)*

Substitution from the BOC chart gives

Xliq(273 K)*1.472 + Xvap(273 K)*6.093 = 1.880 kJ kg^-1 K^-1

This is nearly satisfied with Xliq = .91 and Xvap = .09:

.91 * 1.472 + .09 * 6.093 = 1.888 kJ kg^-1 K^-1

Therefore 9% of the ammonia is in the vapor state at 273 K after an adiabatic expansion from 100% liquid at 298 K.

When the calculation is repeated for adiabatic compression to 100% ammonia vapor at 298 K, the starting point needs to be an ammonia mist consisting of 94% vapor and 6% liquid in equilibrium at 273 K. This is shown by a second calculation similar to the last based on the requirement

$Xliq(273K) * Sliq(273K) + Xvap(273K) * Svap(273K) = Svap(298K).$

This time, *Xliq* is zero at the end of the adiabatic compression, and substitution from the BOC chart should give

$Xliq(273K) * 1.472 + Xvap(273K) * 6.093 = 5.790$ kJ kg^{-1} K^{-1},

where 5.79 kJ kg^{-1} K^{-1} is the vapor phase specific entropy at 298 K. This is satisfied with $Xliq = .06$ and $Xvap = .94$:

$.06 * 1.47 + .94 * 6.09 = 5.81$

To complete the analysis, we need to use the enthalpy data for the liquid and vapor phases from the same table to check that *deltaQh/Th* and *deltaQc/Tc* have the same magnitude, which is the signature characteristic of a Carnot cycle. We can use the enthalpy data since $dH = dQ$ at constant pressure and both evaporation and condensation in the heat exchangers take place at close to constant pressure. For condensation at the higher temperature, the calculation is straightforward. The heat transferred to the air circulating past the heat exchanger coils is:

$deltaQ(298\ K) = Hvap(298) - Hliq(298) = 1627 - 461 = 1166$ kJ kg^{-1},

and

$deltaQ(298\ K)/298\ K = 3.913$ kJ kg^{-1} K^{-1}

For evaporation at the lower temperature the calculation is modified because the entropy calculation just completed shows that the liquid content changes from 94% to 9% during the evaporation process. Therefore, the heat of evaporation for what was initially a kilogram of liquid is reduced to $.94 - .09 = 85\%$ of the per kilogram total heat of evaporation. (In other words, only 85% of each kilogram of refrigerant passing around the cycle is evaporated on its way through the cooling coil, whereas 100% of the warmer vapor is assumed to liquify in the condenser: adiabatic phase changes account for the remainder.) Thus the amount of heat absorbed from the walls of the heat exchanger tubes would be

$deltaQ(273\ K) = (.94 - .09) * (Hvap(273) - Hliq(273))$

$= .85 * (1605\ kJ\ kg^{-1} - 343\ kJ\ kg^{-1}) = 1073\ kJ\ kg^{-1}$,

and

$deltaQ(273\ K)/273 = 1073/273 = 3.930\ kJ\ kg^{-1}\ K^{-1}$,

which is close to the value of 3.913 kJ/kg/K first found for the higher temperature heat exchange and shows that entropy is conserved for the cycle.

To conclude this example (and this little book), the coefficient of performance for this Carnot cycle refrigerator is given by

$COP = deltaQc/(deltaQh - deltaQc)$

In this case,

$COP = 1073/(1166 - 1073) = 11.5$,

so this example unrealistically exceeds expectations based on measurements for an air conditioning system in use and for our analysis of the Brayton cycle air conditioner (COP = 3.6 and 5.9 respectively) partly because adequate allowance has not been made for the range of temperatures to be dealt with on a hot desert day and for the temperature gradients that must exist in the indoor and outdoor heat exchangers (Ananthanarayanan, Ch. 3)

To check this result, we can calculate the COP from the Kelvin temperatures assuming a Carnot cycle:

$COP = Tc/(Th - Tc) = 273/(298 - 273) = 10.9$,

which suggests an uncertainty of about 5% when compared to the first estimate of COP = 11.5.

Conclusion

This book as been written with the needs of residents in hot, dry desert communities in mind. However, when humidity climbs and ice forms on an indoor heat exchanger in the attic of a house, a new set of problems emerges, as P N Ananthanarayanan suggests:

"Where finned [heat exchanger] coils are used, frost formation between the fins will obstruct the air passages and thus affect the quantity of air over the coil. This in turn will reduce the evaporator temperature, leading to thicker frost formation, i.e. a vicious cycle starts. The frost on the evaporator coil also acts as an insulator and retards the heat transfer. With decreased heat-transfer, the evaporator temperature will tend to go down, causing thicker frost and a further decrease in the evaporator temperature and capacity." (Ananthanarayanan, p. 134)

When ice eventually melts, the water may bypass drain systems and cause damage.

Compressor start-up capacitor failure and cooling fan motor failure can shut a system down, but these are relatively easy to spot and repair. Compressor system failure is much more difficult to remedy, and the entire system may need to be replaced because refrigerants once in wide use can no longer be obtained.

Caution is urged when working on air conditioners because of the risks of electric shock and of venting refrigerant to the atmosphere, and it is wise to seek advice.

The End

Works Cited

Ananthanarayanan, P N, *Basic Refrigeration and Air Conditioning.* McGraw Hill Education (India) Private Limited, New Delhi, 2013.

Batchelor, G. K., *An Introduction to Fluid Dynamics.* Cambridge University Press, Cambridge, 1987.

Battino, Rubin, "Mysteries" of the First and Second Laws of Thermodynamics. Chemical Education Today, Vol. 84 No. 5, May 2007.

Bernstein, Jeremy, *Einstein and the existence of atoms*, Am. J. Phys. 74(10), October 2006, pp. 863-872.

Born, Max, *Atomic Physics.* Hafner, New York, Fifth Edition.

Brasz, Joost J., (Carrier Corporation, Syracuse, New York 13221), *Short Course on Throttle Loss Power Recovery in Refrigeration and Cryogenics*, 21st IIR International Conference of Refrigeration, Washington DC, August 17-22, 2003.

Carnot, S, *Reflexions sur la Puissance Motrice du Feu.* Bachelier, Paris, 1824.

Condon, E. U., *Handbook of Physics*, edited by E. U. Condon and Hugh Odishaw. McGraw-Hill, New York, 1958. Part 5 Chapter 1 Principles of Thermodynamics, pp. 5-3 to 5-10.

Derry, T. K., and Trevor I. Williams, *A Short History of Technology.* Dover, New York, 1963.

Emanuel, Kerry, *Hurricanes: Tempests in a greenhouse.* Physics Today, August 2006, Pp. 74-75. www.physicstoday.org

Farey, John, *A Treatise on the Steam Engine.* Longman, Rees, Orme, and Green, 1827. (David & Charles Reprints, etc.)

Feynman, Richard P., Robert B. Leighton, and Matthew Sands, *The Feynman Lectures on Physics Volume I.* Addison Wesley, Reading, 1965.

Garriga y Buach, Josef, and Josef Maria De S. Christobal, *Curso de Quimica General*. Paris, 1804.

Gordon, J. E., *The New Science of Strong Materials*, Princeton University Press, Princeton, 1976.

Haar, Lester, and John S. Gallagher, *Thermodynamic Properties of Ammonia*. Journal of Physical and Chemical Reference Data 7, 635 (1978).

Hays, L. G., and J. J. Brasz, *Two-Phase Turbines for Compressor Energy Recovery* (1996) International Compressor Engineering Conference. Paper 1179. https.//docs.lib.purdue.edu/icec/1179

Hardy, G. H., *A Course of Pure Mathematics*, Cambridge University Press, Cambridge, 1960.

Harrison, David M., *Basal Metabolism-Upscale*-University of Toronto, www. upscale.ca/PVB/Harrison.

Hunter, Louis C., *Steamboats on the Western Rivers*, Dover, New York.

King, Gilbert, *Fritz Haber's Experiments in Life and Death*. Smithsonian Magazine, June 6, 2012.

Leighton, Robert B., *Principles of Modern Physics*, McGraw-Hill, New York, 1959.

Levine, Ira N., *Physical Chemistry*. McGraw-Hill, New York, 1988.

Lienhard, John H., *Engines of Our Ingenuity No.374*: Prison Treadmills, http://www.uh.edu/engines/epi374

Masterton, William L., and Cecile N. Hurley, *Chemistry Principles & Reactions*. Saunders College Publishing, Philadelphia, 1989.

Maxwell, James Clerk, *A Treatise on Electricity and Magnetism, Volume I*. Unabridged Third Edition of 1891, Dover, New York.

Newton, Isaac, *Mathematical Principals Volume One: The Motion of Bodies*, translated by Andrew Motte in 1729 and revised by Florian Cajori. University of California, Berkeley, 1962. Book One, Section I.

Oliveira, Mario J. de, *Exact and inexact differentials in the early development of mechanics and thermodynamics*. Revista Brasileira de Ensino de Fisica, Rev. Bras. Ensino Fis. vol.42 Sao Paulo 2020 Epub Dec 02, 2019, ISSN 1806-9126.

Page, Leigh, *Introduction to Theoretical Physics*. D. Van Nostrand Company, New York, 1935. [Section] 85, Entropy, pp. 288-292.

Planck, Max, *The Theory of Heat Radiation*, translated by Morton Masius. Dover, New York, 1991.

Planck, Max, *Thermodynamics*, translated by Alexander Ogg. Longmans, Green, and Co., London, 1903.

Reitz, Rolf D, (1990), *A Photographic Study of Flash-Boiling Atomization*, Aerosol Science and Technology, 12:3, 561-569, DOI: 10.1080/02786829008959370
https://doi.org/10.1080/02786829008959370

Ripper, William, *Steam Engine Theory and Practice*. Longmans, Green And Co, London, 1914.

Sandfort, John F., *Heat Engines*. Anchor Books, Doubleday & Company, Garden City, 1962.

Saslow, Wayne M., A History of Thermodynamics: The Missing Manual. Entropy (Basel). 2020 Jan; 22(1): 77. Published online 2020 Jan 7. doi: 10.3390/e22010077. Section 12.

Sears, Francis Weston, and Mark W. Zemansky, *University Physics*. Addison-Wesley, Reading, 1955.

Skinner, Anne R., and Dean W. Chandler, Spectroscopy with supersonic jets, Am. J. Phys. 48(1), Jan. 1980.

Stoecker, Wilbert F., *Industrial Refrigeration Handbook*. 1998. [Section] 2.25 Dry Versus Wet Compression. arco-hvac.ir/wp-content/uploads/2018/04/industrial-refrigeration-handbook-1998-wilbertstoeker.pdf

Timoshenko, S, *Theory of Elasticity*. McGraw-Hill, New York, 1934.

Trollope, Fanny (Frances Milton Trollope), *Domestic Manners of the Americans*. Gutenberg.

Watson, E. P., *The Modern Practice of American Machinists & Engineers*. Henry Carey Baird & Co, Philadelphia, 1880.

White, Jack R., *Herschel and the Puzzle of Infrared*. American Scientist, Volume 100, 2012 May-June. www.americanscientist.org

White, Jr., John H., *American Locomotives*. The Johns Hopkins University Press, Baltimore, 1997.

wikipedia en francais, https://fr.wikipedia.org/wiki/Sadi_Carnot_(physicien)

Williams, Richard, *Joseph Black and Latent Heat*. This Month in Physics History. APS News, April 2012. Volume 21, Number 4. https://www.aps.org/publications/apsnews/201204/physics history.cfm

Worthing, Archie G, and David Halliday, *Heat*. John Wiley & Sons, New York, 1948.

Youmans, Edward L., *A Class-Book of Chemistry*. D. Appleton and Company, New York, 1875.